T0228087

NEUROPROSTHETICS
Principles and Applications

REHABILITATION SCIENCE IN PRACTICE SERIES

Series Editors

Marcia J. Scherer, Ph.D.

President
Institute for Matching Person and Technology

Professor
Physical Medicine & Rehabilitation University of Rochester Medical Center

Dave Muller, Ph.D.

Executive
Suffolk New College

Editor-in-Chief
Disability and Rehabilitation

Founding Editor
Aphasiology

Published Titles

Ambient Assisted Living, *Nuno M. Garcia and Joel J.P.C. Rodrigues*

Assistive Technology Assessment Handbook, *edited by Stefano Federici and Marcia J. Scherer*

Assistive Technology for Blindness and Low Vision, *Roberto Manduchi and Sri Kurniawan*

Computer Access for People with Disabilities: A Human Factors Approach, *Richard C. Simpson*

Computer Systems Experiences of Users with and Without Disabilities: An Evaluation Guide for Professionals, *Simone Borsci, Maria Laura Mele, Masaaki Kurosu, and Stefano Federici*

Devices for Mobility and Manipulation for People with Reduced Abilities, *Teodiano Bastos-Filho, Dinesh Kumar, and Sridhar Poosapadi Arjunan*

Multiple Sclerosis Rehabilitation: From Impairment to Participation, *edited by Marcia Finlayson*

Neuroprosthetics: Principles and Applications, *edited by Justin Sanchez*

Paediatric Rehabilitation Engineering: From Disability to Possibility, *edited by Tom Chau and Jillian Fairley*

Quality of Life Technology Handbook, *Richard Schultz*

Rehabilitation Goal Setting: Theory, Practice and Evidence, *edited by Richard J. Siegert and William M. M. Levack*

Rethinking Rehabilitation: Theory and Practice, *edited by Kathryn McPherson, Barbara E. Gibson, and Alain Leplège*

Forthcoming Titles

Cognitive Rehabilitation, *edited by Charles J. Robinson*

Human Computer Interface Technologies for the Motor Impaired, *edited by Dinesh K. Kumar and Sridhar Poosapadi Arjunan*

Neurological Rehabilitation: Spasticity and Contractures in Clinical Practice and Research, *edited by Anand D Pandyan, Hermie J. Hermens, and Bernard A. Conway*

Physical Rehabilitation, *edited by Charles J. Robinson*

Rehabilitation Engineering, Science, and Technology Handbook, *edited by Charles J. Robinson*

NEUROPROSTHETICS
Principles and Applications

Justin C. Sanchez

CRC Press
Taylor & Francis Group
Boca Raton London New York

CRC Press is an imprint of the
Taylor & Francis Group, an **informa** business

CRC Press
Taylor & Francis Group
6000 Broken Sound Parkway NW, Suite 300
Boca Raton, FL 33487-2742

First issued in paperback 2017

© 2016 by Taylor & Francis Group, LLC
CRC Press is an imprint of Taylor & Francis Group, an Informa business

No claim to original U.S. Government works

ISBN-13: 978-1-4665-5323-1 (hbk)
ISBN-13: 978-1-138-74944-3 (pbk)

Visit the Taylor & Francis Web site at
http://www.taylorandfrancis.com

and the CRC Press Web site at
http://www.crcpress.com

To Karen, our lives together have been a remarkable adventure. I have loved sharing it with you.

To Mia, I will always remember the look in your 6-year-old eyes when you said, "Dad, you are really writing a book! Are you finished yet?"

Contents

Preface

How lucky are we to live in a time when it is even conceivable to think that we can build direct interfaces to the brain? And why would we ever endeavor to do so? The answer is quite simple; to repair, replace, and support nervous system function for the human condition. Although what you will read in this text is a collection of the fundamental neuroscience and engineering concepts needed to build neuroprosthetics, it is important to remember that there are ultimately real people for which neuroprosthetics are meant to serve. To the student, scientist, or professional reading this text, go out and meet firsthand those people who could benefit from neurotechnology. It will change your life. Throughout my experience, I have met extraordinary people who are looking for a glimmer of hope to change the quality of their lives. From the mother pleading for help with her child with intractable epilepsy, to the military veteran who lost all of their limbs in service of our country, to the person living with paralysis just looking to reach out and grab ahold of their loved one, it is our brain function that enables life and contains the ingredients that makes us human. When you go about developing neuroprosthetics for your applications, learn from these experiences and think deeply about the implications of interfacing technology directly with the brain. As we look to the future of neuroprosthetics, a fantastic journey into the mind will emerge. What is included in this text is just one small speck in the expanse of this technology. How might a direct neural interface change your life? The possibilities are endless.

Acknowledgments

Science and technology development is by nature a highly collaborative process. Thank you to all those who have contributed to the topics in this text. Your collective insights have helped to advance the field of neuroprosthetics.

Chapter 1: David Clifford, Peter Macdonald, Jeff Kessler, Dave Blakley, and Silvia Vergani

Chapter 5: Abhishek Prasad

Chapter 8: Katie Gant and Scott Roset

Chapter 10: Nicholas Maling and Michael Okun

Author

Justin C. Sanchez is a neurotechnologist, neuroscientist, and neural engineer. He has served as a program manager at DARPA to explore new directions in neural interface technology, brain science, and systems neurobiology. Before joining DARPA, Dr. Sanchez was an associate professor of Biomedical Engineering and Neuroscience at the University of Miami and a faculty member of the Miami Project to Cure Paralysis. He directed the Neuroprosthetics Research Group, where he oversaw the development of neural interface medical treatments and neurotechnology for treating paralysis and stroke, and for deep brain stimulation for movement disorders, Tourette's syndrome, and obsessive–compulsive disorder. Dr. Sanchez has developed new methods for signal analysis and processing techniques for studying the unknown aspects of neural coding and functional neurophysiology. His experience covers in vivo electrophysiology for the brain–machine interface design in animals and humans where he studied the activity of single neurons, local field potentials, and electrocorticogram in the cerebral cortex and from deep brain structures of the motor and limbic system. He has published more than 75 peer-reviewed papers, holds seven patents in neuroprosthetic design, and authored a book on the design of brain–machine interfaces. He has served as a reviewer for the National Institutes of Health (NIH), the Intelligence Advanced Research Projects Activity (IARPA), the Department of Defense's Spinal Cord Injury Research Program, and the Wellcome Trust, and as an associate editor of multiple journals of biomedical engineering and neurophysiology. Dr. Sanchez's degrees include a Doctor of Philosophy, Master of Engineering, and Bachelor of Science, all from the University of Florida, Gainesville, Florida.

Chapter 1 Design thinking for neuroprosthetics

Eventually everything connects—people, ideas, objects. The quality of the connections is the key to quality per se.

Charles Eames

Learning objectives

- Understand the connections between "object centered" and "human centered" neuroprosthetic design.
- Explain the concepts and fundamentals of design thinking.
- Master the tools of design thinking such that they can be applied to any neuroprosthetic application.

1.1 Introduction

The goal of this text is to equip readers with the fundamental knowledge needed to innovate in the field of neuroprosthetics. It is aimed at a broad audience that consists of undergraduate and graduate students, professionals interested in neuroprosthetics, and even the non-scientist who would like to dig deeper into the concept of a neural interface. At their core, neuroprosthetics are remarkable inventions because they fundamentally change the way that biology expresses itself. They are engineered devices that provide an alternative communication channel for the way that the brain interacts with itself, the body, and the environment. Neuroprosthetics are "human centered" because they address people's needs as they interact with the world during their daily lives. Depending on the person, those needs could be restorative as in the loss of function after injury and disease, or they could be supportive in facilitating communication and control of devices for the able bodied.

At the time of this first publishing (2013–2014), neuroprosthetics are at a point where they are coming of age. The field has moved beyond the initial proof of concept of the potential to tap into neural circuits and use the signals to do work with the external environment. Those initial proofs of concepts have sparked an international push toward new investments in brain science, and innovative neuro-technologies have the potential to unify understanding of brain and open the use of neural interfaces for a wide variety of medical and nonmedical uses (Shen, 2013).

Throughout the history of neuroprosthetic development, federal agencies have supported highly innovative research by fostering multidisciplinary collaborations among neuroscientists, physicians, mathematicians, and engineers who have had the goal to address four major challenges: detect—develop diagnostics, models, and devices to characterize and mitigate threats to human brain; emulate—leverage inspiration from functional brain networks to efficiently synthesize information; restore—reestablish behavioral and cognitive function lost as a result of injury to the brain or body; and improve—develop brain-in-the-loop systems to accelerate training and improve functional behaviors (Miranda et al., 2015).

It is important to note that in this early stage of this text, the study of complex systems like neuroprosthetics typically utilizes a divide-and-conquer approach where the focus is on brain itself and it is viewed as an "object" that can be broken into many manageable pieces for study. This is illustrated by the foundational work of Santiago Ramon y Cajal and Raphael Lorente de No, who initiated approaches to brain science in the domain of segmentation and taxonomy (Witkowski, 1992), as well as David Hubel and Thorsten Wiesel, who made discoveries through advanced electrode technology to study brain activity in the visual cortex (Hubel and Wiesel, 1959). This fundamental "object centric" approach to partitioning the brain into its motor, visual, auditory, somatosensory, and cognitive components and studying them in exquisite detail has been repeated in countless laboratories internationally and forms the basis for much of the neuroscientific knowledge about how the brain and neuroprosthetics should be designed.

As we embark on the 21st century efforts to develop neural interfaces, the fragmented "object centric" approach for brain research related to neuroprosthetic design is no longer sufficient to support the next generation of growth. Today, neuroprosthetics are not about the components but about the function that they deliver to their users. As the technological barriers to studying the brain as a whole are being lowered, it is unavoidable in the not-too-distant future that many subsystems of the brain will be accessible in real time with potentially better resolution than that could be imagined by Hubel and Wiesel. Accessibility to those structures opens up more facets of neuroprosthetic function. The focus is on how users leverage the neuroprosthetic to support how they work, interact, and play. This shift from "object centered" to "human centered" neuroprosthetic design changes the way that one should approach the science and technology of neural interface development and is the focus of the next sections.

1.2 Design thinking

The concept of design thinking provides a framework to base the shift from "object centered" to "human centered" neuroprosthetic design (Kelley, 2007). A core value in human-centered design is exploring and understanding people's needs. Neuroprosthetic users have very diverse needs depending on their physical condition, neurophysiologic and neuropsychiatric state, personal relationships, ability to care for themselves or care for others, or responsibilities in daily life activities. The main objectives of neuroprosthetic design thinking is to identify opportunity areas where neurotechnology can bring differentiating value for users. This involves identifying concepts and features of devices that resonate well with user's needs. Design research is very different from traditional academic research because it is a search for inspiration to gain empathy and knowledge about the stakeholders the neuroprosthetic system is being designed for. It often involves forgetting assumptions and approaching scientific, engineering, and applications of technology from a new perspective. The approach of design thinking is often qualitative but is based in facts about user's needs, desires, habits, attitudes, and experiences with neuroprosthetics. Those facts can then be transformed into quantitative neuroscience or engineering specifications to guide research development. Table 1.1 highlights several of the major differences between the traditional neuroprosthetic research and the design-thinking approach to arrive at new neuroprosthetic principles.

Neuroprosthetic design thinking is initially generative as the process explores possible directions, ideas, and concepts and evolves as new knowledge about users is obtained. Latter phases are convergent and about evaluating ideas, ranking them, filtering some out, and focusing on others to develop refined designs.

Table 1.1 Comparison of Traditional Neuroprosthetic Research with Neuroprosthetic Design Research

Traditional Neuroprosthetic Research	Neuroprosthetic Design Research
Improves existing knowledge about the brain and proves hypotheses	Inspires new ideas about neuroprosthetic use
Criticizes ideas about neuroprosthetics to make them stronger	Defers judgment and builds on others' neuroprosthetic concepts to make them stronger
Uses well-controlled laboratory experimental paradigms	Visits natural contexts to understand neuroprosthetics in daily life
Involves scripted, fixed protocols in the study of neuroprosthetics	Uses dynamic interaction and study of neuroprosthetic use
Looks for large patterns of brain behaviors in the mean and statistical sense	Learns from extreme users to take a look into future neuroprosthetic uses
Focuses on individual's contribution to neuroprosthetic research	Is a highly collaborative process among many neuroprosthetic stakeholders

3

It often involves combining that knowledge with information about competing technologies, their current use, and their future potential. For example, consider a person living with paralysis and the myriad of movement assistive technologies like eye trackers, sip and puff controllers, powered wheelchairs, and so on, that are available to them today. It is important to understand these technologies, products, or tools in terms of what users utilize the most. Asking questions such as "What are the major challenges associated with using them or having their caregivers use them?" provides real-world context to the use of such systems. Their use leads to the value proposition associated with different types of neuroprosthetic solutions. Ultimately, designers need to know about the key experience features that appeal to users and what are the behaviors, attitudes, and needs relevant for understanding users as they first encounter and use the neuroprosthetic.

1.3 Inspiration for neuroprosthetic design

Where does inspiration for neuroprosthetic design come from? Are there "lone geniuses" out there formulating how to assemble fundamental neuroscience with engineering for some "killer application?" The answer is a resounding "No" because neuroprosthetics are complex systems that involve highly specialized expertise from a variety of fields, which need to be aggregated in an intelligent way to produce a function for the nervous system. Function is the key in this discussion because users are seeking to use the neuroprosthetic to do something. Evoking knowledge from users can often be challenging, and there are a variety of inspiration techniques that can be used to stimulate the process. The five activities neuroprosthetic designers should implement before defining specifications for new neural interface systems are as follows:

1. Draw yourself in 1 year: This exercise inspires users to project themselves out into the future to envision how they could potentially be changed by neurotechnology. It inspires creativity to contrast how users view themselves today with the version of themselves that they hope to be. An example drawing is shown in Figure 1.1a.
2. Need card prioritization: Need cards are another approach to seeding a user's concepts through the identification of specific traits or attributes that are essential for neuroprosthetic design. By decomposing the neuroprosthetic system into its components, such as fully implantable, wireless, autonomous, and so on, and by asking the user to rank them, the interviewer can assess the priorities for neuroprosthetic design. An example ranking is shown in Figure 1.1b.
3. Sacrificial concepts: This exercise seeds the interviewee's ideas of neuroprosthetics with the interviewer's concepts as shown in Figure 1.1c. This could be 5–10 neuroprosthetic designs that have a particular function, a

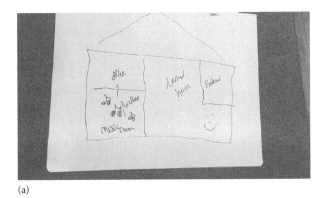

(a)

(b)

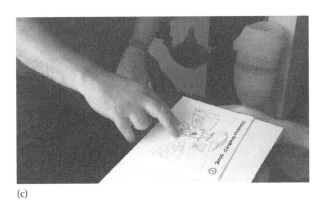

(c)

Figure 1.1 Examples of techniques used to inspire neuroprosthetic design.
(a) Draw yourself in 1 year, (b) need card prioritization, and (c) sacrificial concepts.
(*Continued*)

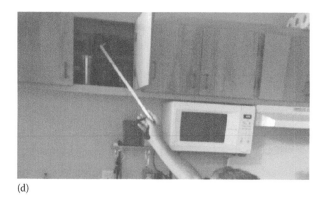

(d)

Figure 1.1 (Continued) Examples of techniques used to inspire neuroprosthetic design. (d) in-context observation.

form factor, or a use that a user would be asked to comment on. It is a method to see what designs resonate well with a user's needs.

4. In-context observations: One of the best ways to learn about function is through in-context observations. This is not a sterile laboratory setting but the real environments where neuroprosthetics likely appear. This could be somebody's home, office, or their favorite place of leisure, as shown in Figure 1.1d. Often in those environments, great insight can be gained because what users say they do is not always what they actually do.

5. "How might we?": This phrase can be used as a stimulus tool to evoke novel solutions to user's needs. For example, "How might we design a neuroprosthetic for users' actual priorities, not just moving their arm again?"

1.4 Prototypical example of neuroprosthetic design thinking

To illustrate the process of design thinking for neuroprosthetics, consider a hypothetical design challenge of building a next-generation consumer product that will enable a person to actuate devices in their home using only their neural signals. Such a challenge evokes many questions. What would the product do? What would it look like? How would it be used? Who would be best suited to use a device like this? These questions can be answered using a variety of techniques. They include communicating with experts who may have insights to this problem or related problems and using inspiration tools described in Section 1.3 to identify specific attributes of the potential product. Ultimately, it is the synthesis of all of the information collected from experts and inspirational activities that leads to the development of a design concept for the actual device to be built.

1.4.1 Expert interviews

A first step in collecting information is to interview experts who are able to share their experience in the domain of interest. In this example, the application space is the home and the people living in it. These people could be able-bodied, or they could be living with an injury to their nervous system. A consideration for if the people are disabled is if they are living with a caregiver or not. Deeply understanding the role of the caregiver and the challenges they face could provide insight to the needs of the patient. Focus on the actual environment (i.e., the home) is critically important. Are there special design features of homes that make them accessible or more conducive to daily life activities? Are there people working on homes of the future that have new design features that would be relevant for a brain-actuated home? From this brief framing of the example problem, a list of candidate expert is provided. This list is not exhaustive and can be expanded or contracted depending on what is learned from the interviews, as follows:

- Registered nurse for a caregiving agency
- Architect or home designer
- Neuroscientist
- Director of communication for the ALS Association
- Director of a human engineering research laboratory
- PhD scientist from a neural engineering laboratory
- Expert in biomechatronics
- Synthetic neurobiologist
- Designer of virtual and immersive environments
- Futurist or technologist

Interacting with a group of people such as the ones described earlier can produce key learning about the problem of actuating the home with neural activity. Interviewees often make comments about their perspective into the problem. These should be recorded and be used to inspire new concepts about the design. At the end of the expert interview process, the designer should distill the information down into a list of key insights and their supporting concepts as in the following:

Neuroprosthetics in the home focus on a sense of purpose

- It is not about recovering a lost function. It is about a user's daily life.
- Motivators are highly personal.
- Engaging the creative mind supports healthy behavior.
- Users perceive a renewed sense of purpose when they shift their focus from self to others.

7

Neuroprosthetics in the home focus on a sense of independence

- A sense of independence comes from the ability to shape one's environment through technology.
- Caregivers look up at the opportunity to shift their role from functional to emotional supporters.

1.4.2 User observations

While the experts may be able to provide technical or generalized insight into the device for actuating a home with your brain, it is the potential users of the device that provide clarity on the highly personalized needs. Given that a brain-actuated home does not necessarily exist today, how does one go about finding the most relevant people to observe? The approach is similar to that of identifying the experts. Seeking out people that have experienced elements of the potential device, people that have extremes in their needs, and people that are living with conditions may be served by the potential device. These could include the following:

- An amputee
- A high-level gamer
- A person diagnosed with PTSD
- A person living with a cochlear implant
- An entrepreneur who works primarily from their home office

All of the aforementioned people have unique ways that they interact with their home and can provide insight to how a neuroprosthetic would potentially affect their current life and their future life. Inspiration tools such as sacrificial concepts and needs priority lists are then provided to stimulate user feedback. Two example sacrificial concepts are provided in the following sections to illustrate how the technique is used in practice.

1.4.2.1 The assisted kitchen   The assisted kitchen enables your home to ask you what you want to eat for a meal, read your mind, prepare the meal, and then clean up afterward. An example of such a kitchen is shown in Figure 1.2a.

User feedback:

I like this...
- "I think of what I want, then it just pops into existence."
- "Reminds me of the Jetsons."
- "Cooking takes me a long time, so I don't really do it."

But...
- "I don't always know what I want. Give me choices."
- "But I love to cook. I don't want to just sit around."
- "I'd rather ask... than have it read my mind."

8

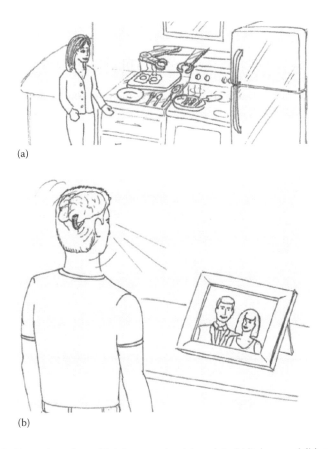

(a)

(b)

Figure 1.2 Examples of sacrificial concepts: (a) assisted kitchen and (b) memory assist.

- "I'm not scared of robots, but having the house interface with my nervous system might be scary."
- "My girlfriend cooks or prepares food ahead for the week, so we wouldn't need this."

Yes, and...

- "Robots would be 'good for simple tasks' like changing dressings, dispensing meds."
- "This may be good for laundry."
- "I'm a big fan of the hands-free toileting experience."

1.4.2.2 Memory assist Memory assist is a brain implant that, when triggered by what you are looking at, will stimulate the appropriate part of your brain to help recover lost memories. An example of such a device is shown in Figure 1.2b.

9

User feedback:

I like this…
- "There's a vet I know with severe TBI and associated memory loss. This would be for him."
- "I'd want one now, if it was safe to help me stay positive, and shut off the bad PTSD."

But…
- "I'd be hesitant to go to the brain unless there were huge benefits."

Figure 1.3 Synthesis of information collected from interviews and observations to arrive at design principles.

1.4.3 Synthesis

After expert interviews and user observations are completed, the next step is to perform synthesis of the information. This consists of aggregating all of the concepts together and looking for common trends or themes in the responses. A stimulating way to perform this process is to have the design team layout all of the insights with color-coded Post-it notes, as shown in Figure 1.3. Rapidly writing down the ideas in stream of consciousness form and organizing them according commonalities and differences will help in the visualization of the core principles. The result of the synthesis process is to arrive at a list of design principles that guide the development of the next-generation device. With the list of design principles, the inspiration technique of "How might we?" can then be used to provide direction to brainstorming sessions that give rise to the actual design of the system. A set of "How might we?" examples is provided as follows:

How might we (HMW)...

1. HMW design for users' actual priorities, not just for their particular disability?
2. HMW design holistically to support and supplement brain actuation of the home?
3. HMW empower users by empowering their caregivers?
4. HMW design a home that trains users to rehabilitate themselves?
5. HMW design the aspirational home, rather than a home for the disabled?

Exercise

1. Design a next-generation neuroprosthetic device. Perform expert interviews and user observations. Use inspiration exercises, including "Draw yourself in 1-year," sacrificial concepts, need card prioritization, in-context observation, and "How might we?" Perform synthesis of the information and report on the core design principles.

Chapter 2 Interfaces to the brain

I've personally reached the point where the sound of MP3s are so uncompelling, because so much is lost in translation.

Beck

Learning objectives

- Understand the fundamentals of how electrodes translate neural activity into signals that are useable by machines.
- State the signal sensing capabilities and trade-offs for common electrode designs.
- Determine the most appropriate electrode for any neuroprosthetic application.

2.1 Introduction

The design of neuroprosthetic devices begins with an interface to the brain. This interface is very different than what nature provides in terms of communication between the nervous system and the body because, for neuroprosthetics, brain activity needs to be machine-readable. Today, machines, computers, and virtually any other engineered system are fundamentally unable to directly interpret the physiological processes that occur in the nervous system. Therefore, those processes whether they are electrical, chemical, or physical in nature must be translated from their natural domain into signals that can be used by analog or digital hardware. The "neuron doctrine" principally states that groups of single neurons give rise to all brain functions (Witkowski, 1992). Those neurons operate through electrochemical processes that cause them to fire and produce action potentials, which is the signaling output of the cell. Any process related to the generation and time-varying production of action potentials can be used as a control signal for a neuroprosthetic device provided that those signals are related to the primary function of interest (i.e., movement, memory, decision-making). Processes of interest could be the change in extracellular dendritic ionic currents, depolarization of the neuron itself, blood flow supporting neuron metabolism, or the acoustic signature of an action potential traveling down an axon. Whatever the choice of signal, there are multiple requirements that are necessary for real neuroprosthetics, which function in everyday life for humans. The interface must be capable of producing real-time representations of neuronal activity, it must be reliable and generate repeatable signals, and the interface must be reasonably portable. In this chapter, the fundamentals of common electrophysiological neural interfaces for neuroprosthetics are presented.

2.2 Electrical interfaces

To capture the physiological processes associated with activity in neural tissue, an interface needs to be formed between the brain and the electronic device used in the neuroprosthetic system. Neurons are electrochemical cells that polarize and depolarize themselves with respect to their extracellular environment. Through the generation of action potentials, neurons change their resting potential from approximately -70 to 55 mV by exchanging ions from their internal environment to the external environment (Kandel et al., 2000). Ions such as sodium (Na^+), potassium (K^+), and chloride (Cl^-) are involved in the potential change as they traverse the neuron's permeable membrane through voltage-gated ion channels to produce the action potential. The production of action potentials creates an ever-changing flow of ions in the internal and extracellular spaces of brain tissue and this change in ions is a direct representation of ongoing neural activity and representation. The sole purpose of building an electrode interface with the brain is to change the aforementioned ionic current into electronic current that can flow in a neuroprosthetic device. The transformation is mediated by the interface that is at the intersection of the electrode and electrolyte or, in more specific terms, the physical electrode and tissue interface. The role of the interface is to conduct current across it. This current is carried by ions in the biotic component (neural tissue) of the interface and by electrons in the electronics of the neuroprosthetic. In the next section, the chemistry associated with the reactions at this interface will be presented.

2.2.1 The interface problem

Figure 2.1 shows a detailed schematic of the neural electrode interface, which represents the initial location of data exchange between the brain and the neuroprosthetic system electronics. In this example, the metal that makes up the electrode recording site is represented as atoms C. Note that the electrode recording site is typically of a fixed surface area of either a round, rectangular, or conical geometry. Common electrode materials include platinum (Pt), iridium (Ir), tungsten (W), and alloys of platinum and iridium (Pr-Ir). Although the electrode recording site metal is exposed to the electrolyte, the rest of the electrode is insulated and

14

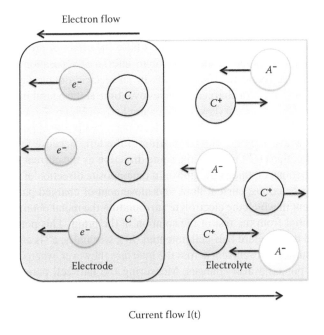

Figure 2.1 The neural electrode interface.

protected from contacting the electrolytic solution. A constant recording surface area is important for maintaining the impedance. The impedance takes the form $R = \dfrac{\rho L}{A}$ where R is the impedance, ρ is the resistivity, L is the length of the conductor, and A is the surface area. As the electrode area increases the impedance decreases and vice versa. When the metal comes in contact with the electrolyte, the surface area is exposed to an aqueous solution containing cations of the electrode metal C^+ and anions A^-. A series of oxidation and reduction reactions occur at the interface.

2.2.2 Half-cell potential

The chemical reactions that govern the exchange of ions at the electrode–tissue interface are given by $C \rightarrow C^+ + e^-$ and $A^- \leftarrow A + e^-$. Here, when the metal C contacts the electrolytic electrolyte fluid of the brain, the aforementioned reactions are triggered. In this example, the metal C is oxidized and gives off an electron whereas the anion A^- gains an electron to be reduced. The chemical reaction begins immediately and, as a result, local concentration of cations at the surface of the electrode recording site changes. As described previously, the surface area not only plays a role in the impedance but it, in a related way, also affects the oxidation and reduction reactions. The larger the surface area, the greater the effect

15

the reaction has on the system and the exchange of charged particles can lower the impedance. Smaller surface areas have the opposite effect and thus larger impedance. The oxidation and reduction reactions have the effect of producing charge in many regions of the electrode–electrolyte interface. Separation of charge at the electrode–electrolyte interface results in an electric double layer (bilayer). Tracking the movement of charged particles over time at the neural interface gives rise to the concept of current flow. In the example from Figure 2.1, current moves from left to right as a function of time. Electrons move in the opposite direction to current flow and they are shown moving to the left. Because the metal C is oxidized, the cations (C^+) move in the same direction as the current flow. In contrast, the reductions of anions (A^-) move in the opposite direction of current flow. The chemical reactions, current flow, and movement of charged particles create an environment in which the electrolyte surrounding the metal obtains a different electric potential from the rest of the solution in the brain. This potential difference is called the standard half-cell potential (E^0). Normally, E^0 is an equilibrium value and assumes zero-current across the interface; however, when current flows, the half-cell potential also changes. Measuring the half-cell potential requires the use of a second reference electrode and by convention a hydrogen electrode is usually chosen as the reference. By definition to capture this value, hydrogen is bubbled over a platinum electrode and the potential is defined as zero. The Nernst equation governs the half-cell potential and is given as $E = E^0 + \dfrac{RT}{nF} \ln\left(a_{C^{n+}}\right)$ where E is the half-cell potential, E^0 is the standard half-cell potential, R is the universal gas constant (8.31 J/(mol K)), T is the absolute temperature in K, n is the valence of the electrode material, F is the Faraday constant (96,500 C/(mol/valence)), and a is the ionic activity of cation C^{n+}.

2.2.3 Electrode polarization

The efficiency of a neural interface to complete its oxidation and reduction reactions depends on the thermodynamically determined potential for each of the half reactions to naturally occur and its relationship to the actual potential at which the reactions actually take place experimentally. A number of influencing factors can cause these oxidation and reduction reactions to deviate from their theoretical values. Cumulatively, the sum of the potentials above the theoretical value is called the overpotential (V_p). For neural tissue interfaces, the major components of the overpotential include the ohmic potential (V_r), which is caused by the resistance of the electrolyte, the concentration potential (V_c), which is caused by the redistribution of the ions in the vicinity of the electrode–electrolyte interface (concentration changes), and the activation potential (V_a), which is caused by the activation energy barrier for metal ions going into solution. Depending on the electrode metal type and the existence (or absence) of an overpotential, neural interface electrodes are classified as either being perfectly polarizable or

16

nonpolarizable. Polarizable and nonpolarizable electrodes have unique properties that make them each well suited for neuroprosthetic applications. Perfectly polarizable electrodes are commonly made from noble metals such as platinum. They are difficult to oxidize and no charges cross the electrode when current is applied. These are well suited for stimulation applications because they do not dissolve. Their main mechanism of action is just changes in concentration of ions as the interface. They behave, in general, like capacitors. For nonpolarizable electrodes, such as silver/silver-chloride electrodes, all charge freely crosses that electrode–tissue interface when current is applied. No overpotential is generated and these electrodes generally behave as resistors.

2.2.4 Electrode circuit model

An equivalent electrical-engineering circuit can be used to represent the combined properties of neural electrode interfaces. Figure 2.2 shows the design and consists of a battery connected to a resistor and capacitor in parallel, which are then connected to a resistor in series. Here, the battery represents the half-cell potential (E_{hc}) of the oxidation and reduction reactions. The electric double layer that results from perfectly polarizable electrodes can be represented by the capacitor C_d. For nonpolarizable electrode properties, R_d is resistance to current flow across the electrode–electrolyte interface. Lastly, R_s is the resistance associated with the conductivity of the electrolyte.

2.2.5 Noise sources

Electrodes that are used in neuroprosthetic applications are subject to intrinsic and extrinsic noise. Intrinsic noise is due to the electrode materials or reactions that

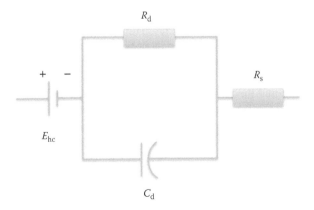

Figure 2.2 Equivalent circuit model for neural electrode interfaces.

17

occur at the electrode–tissue interface. Extrinsic noise is a result of sources external to the electrode. These could be unwanted signals from other equipment in the room, 60 Hz power line interference from fluorescent lamps, or biological noise such as muscle artifacts. External noise is commonly handled through a variety of techniques, which include insulation, isolation, or adaptive noise-canceling techniques. Intrinsic noise is more difficult to contend with and may often be out of the control of the experimenter. These include thermal noise, which is also known as Johnson or Nyquist noise and is due to the random motion of the charge carriers, which corresponds to the resistive component of the electrode impedance. The voltage associated with thermal noise takes the form $v_n = \sqrt{(4k_B TR\Delta f)}$ where k is the Boltzmann's constant (1.38×10^{-23} J/K), T is the absolute temperature (293 K), R is the resistance, and f is the bandwidth for the neural signal of interest. From the equation, the noise increases with temperature and electrode impedance because it is due to the random movement of charge carriers. Like changes in temperature and impedance, random fluctuations of the electrode half-cell potentials can introduce noise. Half-cell potential changes can be caused by electrode surface contamination or inconsistencies in the electrode material. These variations directly translate into changes in the capacitance In general, any disturbance of the electrode–electrolyte double layer can result in shifts and instabilities in the double layer capacitance. Two other forms of noise commonly analyzed in neural electrode interfaces include Schottky (Shot noise) and flicker noise ($1/f$ noise). Schottky noise is due to charge passing over energy barriers and flicker noise is intrinsic in physical electronic systems.

2.2.6 Tricks of the trade

There are a multitude of good experimental practices that can facilitate the acquisition of high-quality neural signals in practice. Deviation from these practices can lead to degradation in the signal-to-noise ratio and stability of neural signals used in neuroprosthetic applications. The first priority in electrode usage is to ensure the integrity of the acquisition system. This includes making sure all solder joints in all of the connectors between the neural electrode and the data acquisition system are good. If the movement of the electrode or cabling system introduces noise into the recordings, then there is a problem with the connector. A new connector needs to be resoldered to fix the problem. Strain relief should be added to any electrodes or cabling that is in movement or under tension. This can be achieved by creating gradual mechanical transitions in the wire that absorb forces being placed on the wire. Repeated flexing of cables can break internal wires and thus adding either mechanical stabilization or extra insulation can prevent against breakage. After fixing any weak connections and providing strain relief, the insulation on the neural electrode should be checked. All portions of the electrode and cabling should have intact insulation. The only exposed portion is the actual recording site of the electrode that is in contact with the neural tissue interface. Electrode insulation should be periodically checked because externalized cables can become cracked or cut during use. In addition, electrodes with polymer insulation that are chronically exposed to extracellular fluid can degrade over time. This can come in the form of micro cracks along the shank of electrodes or delamination of the insulation around the edges of the recording site. Some polymers such as polyimide are also known to absorb water. If this does occur, the insulation becomes a part of the electrode and the effective resistance decreases because of the increased surface area. Large area, low impedance electrodes are more susceptible to extraneous noise. The problem of insulation can be compounded if the components of the electrode in contact with the electrolyte are made of different materials. Because each metal has different half-cell potentials, noise can be introduced if they are in contact together in the presence of an electrolyte. This recommendation also applies to differential measurements that are made with two different electrodes. Investigators should seek to make each electrode the same material to minimize the effects of noise. Lastly, it is important to appropriately match recording electronics to the neural electrode that is chosen. In most electrodes used for neuroprosthetic applications, voltages are being measured from neurons. In this scenario, it is desirable for the input impedance for amplifier systems to be as large as possible compared to the impedance of the electrode. With this impedance difference, the amplifier will not draw any current at its input and thus will not affect the voltages to be measured for the neuroprosthetic.

2.3 Electrode design

Designers of electrodes for neuroprosthetic applications seek to optimize specifications of sensitivity, selectivity, and safety/reliability (Wolpaw and Wolpaw, 2011). Fundamentally neural electrodes need to be sensitive, which means that they must transduce sufficient information from neural tissue to support the function of a neuroprosthetic system. As described previously, the transduction begins with the exchange of charged particles at the electrode–tissue interface but this signal needs to be abstracted up to the hierarchical organization of the nervous system that contains neurons, networks, systems, and ultimately cognitive representation (Sejnowski and Churchland, 1989). Selectivity refers to the ability to record or stimulate one, tens, hundreds, or thousands of neurons in that hierarchy with any combination and at any time resolution. With electrode systems that are selective comes the ability to correlate the activity of neurons or neural systems with motor, sensory, and cognitive (neuropsychiatric, decision making, perceptual) function. Potential uses for selective electrode interfaces include neuroprosthetics for movement, vision, memory, and deep brain stimulation (DBS) for neuropsychiatric conditions. The third major design consideration is safety/reliability. Neuroprosthetic electrodes are ultimately designed for humans and therefore cannot do harm to humans. Safety profiles must make sense for the benefit–risk ratio. This means that the level of function that is obtained from the neuroprosthetic system needs to be weighed against the risk of placing electrodes in the nervous system in the first place. The degree of invasiveness must also not exceed what is absolutely necessary for the function that is provided. The benefits and risks are also often expressed in terms of the reliability or how long the electrode interface will continue to produce viable, consistent signals during the time that they are in contact with the nervous system. It would not be uncommon for a neuroprosthetic system to have to maintain function for decades after being implanted. To achieve the previously mentioned design specifications, electrode fabricators must define the functional components that the device uses to transduce signals from the nervous system. This begins with determining the specification of the electrode sites themselves that would be used for recording or stimulating. Important factors include the electrode size, material, surface area, roughness, and shape. The overall shape of the electrode can play a large role in the insertion of the interface into the neural tissue. Sharpened or conical-tipped electrodes tend to pierce neural tissue with less damage compared to blunt-tipped electrodes. The electrode size plays a role in the ability to localize and detect neuronal sources. The smaller the recording site, the more localized the recording becomes due to the effect of impedance. Large electrodes have low impedance and can capture broader neural activations. Typically, microelectrodes for recording single neuron activity are 50 μm in diameter (or smaller) whereas macroelectrodes for recording the electroencephalogram can be as large as 1 cm. Figure 2.3 shows the spatial domains for single neuron action potentials, local field potentials (LFP), the electrocorticogram (ECoG), and electroencephalogram (EEG). The biophysics of electrical signals in brain tissues governs the fall off of potential as a function of the distance between the electrode and the neural source (Nunez, 1981). This relationship is $\frac{1}{\sqrt{d}}$ where d is the distance. Because the distance between an EEG electrode and a single neuron source is so large, the ability to detect a single action potential from that modality is greatly diminished. The electroencephalogram is generated from the synchronous activity of large populations of neural ensembles that spans many centimeters of the cortex. The orientation of the fields that are generated by action potentials also plays a role in signal acquisition. Neuron electric fields can be approximated by dipoles that are aligned with the long axis of the neuron. Any variations in the alignment of those dipoles due to the natural convexities of cytoarchitecture can affect the spatial summation of the potentials that the electrode can detect. This fundamental relationship between aggregate activation of neural sources, electrode physical properties, and biophysics affects the spatial specificity across

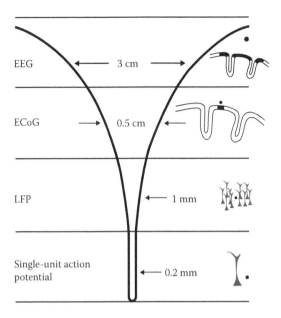

EEG ←— 3 cm —→

ECoG —→ 0.5 cm ←—

LFP ←— 1 mm

Single-unit action potential ←— 0.2 mm

Figure 2.3 Comparison of the spatial domains of four electrical interfaces for neuroprosthetics. (Adapted from Schwartz, A., X. Cui, D. Weber, and D. Moran, Brain-controlled interfaces: Movement restoration with neural prosthetics, *Neuron*, 52(1), 205–20, 2006.)

the hierarchy of single neurons, small neural populations (LFP), mesoscopic neurodynamics (ECoG), and macroscopic neural activity (EEG).

The material, surface area, and roughness have a coupled role in the production of high-quality recording and stimulation. Inert materials like platinum are ideal for stimulating and recording due to the noise characteristics associated with the half-cell potential as well its stability when driven with current when used as a stimulator. The selection of materials also plays a role in the ability to fabricate fine wires of small diameter. The Young's modulus of pure platinum or pure iridium make them difficult to form into fine wires. However, alloys of platinum and iridium have optimal mechanical as well as electrochemical properties. Alloys and coatings can also be used to affect the surface area and roughness of electrode recording sites. Rough surfaces have been used to increase the surface area of electrodes while maintaining an equivalent geometry. This has the effect of lowering the impedance while also promoting neural tissue interaction as the surface (Raffa et al., 2007).

Once the electrode recording site specifications have been determined, the next step is to identify how many recording sites are desired and how they should be arranged (two-dimensional [2-D] or three-dimensional [3-D]). The selection of the number and geometry of the electrodes depends on the sites that are being recorded in the nervous system. When electrode recording sites are arranged into an array, they should conform to the cytoarchitecture of interest. For example, for motor neuroprosthetics, it is often desirable to target layer V pyramidal cells in the cortex. These neurons are arranged in a 2-D plane approximately 1.6 mm below the surface of the cortex. To acquire these signals, a 2-D array of micro-electrodes with recording sites at the tips would be appropriate to acquire the signals. In contrast, if the goal is to capture the neuronal processing traversing layers I to VI of the cortex, it would be desirable to have a 3-D array that consisted of multiple microelectrodes, each with recording sites at the tips and along the shanks. This way, large coverage of the 3-D volume of the cortex can be captured.

The next step beyond choosing a recording site specification and array geometry is to design the leads or traces that electrically connects the recording site to the back end electronics. There is a need to select leads that have low resistance to minimize signal loss when the neural potentials are routed. They should have sufficient flexibility as to not introduce external forces on the brain (Thelin et al., 2011). It is desirable to not create any tissue disruption or damage from the tethering of the electrodes to the recording or stimulating neuroprosthetic device. Investigators have used the terms "floating" and "fixed" arrays to describe meth-ods of tethering. Floating arrays essentially float with the brain inside the skull because they have minimal tethering forces applied to them. In contrast, fixed electrodes are mechanically coupled to the skull and have maximum tethering forces applied to them. One of the major factors that influences the tethering forces from the traces is the selection of the material that insulates the wires. Each insulating material, or dielectric, has its associated mechanical stiffness and degree to which it degrades over time. Common polymers for insulating neu-ral electrode interfaces include polyimide, parylene C, and polydimethylsiloxane (Hassler et al., 2010).

2.3.1 Examples of neuroprosthetic electrode arrays

There are a wide variety of electrode array types that can be used in neuropros-thetic applications. The selection of the type to be used depends highly on the application, location of implantation, subject (animal vs. human), and duration of use. The primary goal when selecting electrodes is to acquire the highest-quality signals for the longest duration while preventing any injury or disruption in the neural tissue. The following is an overview of common electrode types.

2.3.1.1 Microwire arrays Microwire electrodes are some of the earliest forms of technology that provided the foundation for neuroprosthetic research. Pioneered by David Hubel in 1957, the basic form of the technology has not changed drastically compared to what is used today. An example of these early electrodes is shown in Figure 2.4a. These electrodes are made of tungsten and are sharpened such that they can more easily penetrate the cortex of the brain. Unlike the electrodes of today that are insulated with biocompatible materials, the early electrodes from Hubel were coated in lacquer. Many of the electrode designs that are used in human neurosurgery such as those that are manufactured by Fred Haer Corporation (FHC) also have very similar design. They are long filaments of tungsten, stainless steel, or platinum/iridium wire that can be packaged and sterilized for human applications. Beyond single filaments, microwires can be assembled into arrays. An example of a tungsten microwire array of 16 electrodes (2 × 8) from Tucker-Davis Technologies (TDT, Alachua, FL) is shown in Figure 2.4b. Here, the wires are connected to a zero insertion force (ZIF) connector. This connection provides a compact and easy way to attach many electrode wires to the subject. The connector just clips onto the metal shroud and provides the contacts back to the headstage and amplifier system for recording. Both the single filament wires as well as the arrays shown in Figure 2.4a and b are "fixed" arrays because when they are chronically implanted, they are rigidly connected to the skull. In contrast, the electrodes that are shown in Figure 2.4c from MicroProbes are "floating" arrays because the array assembly contains an ultraflexible cable routed to the connector that connects back to the recording electronics. This cable allows the array to move in conjunction with the brain. An additional feature of this array is that the electrodes are all cut at different lengths to provide multiple depths for interfacing with the nervous system. Microwire arrays such as the ones described here often have complex fabrication methodologies. There can be steps in the fabrication procedure that involve hand-made components that require alignment, soldering, and cutting. A variety of robotic jigs and laser-cutting techniques have been developed but they are still labor-intensive (Williams et al., 1999).

2.3.1.2 Planar micromachined Advances in electrode array fabrication can be achieved by borrowing techniques from the microfabrication industry (Kipke et al., 2003). Moreover, integration of signal processing circuitry onto the substrate of the electrode is possible and may be desirable to reduce noise pickup. One of the limitations of traditional microwire electrodes is that they only contain one recording site at the tip and therefore can only record neuronal activity at one depth in the neural tissue. To overcome this issue, planar microfabricated electrodes such as the array in Figure 2.5a from NeuroNexus can contain multiple recording sites positioned along the shank of the electrode. Shown by the linear arrangement of "dots" in the figure, the recording sites typically consist of exposed metal pads that are connected via interconnect traces to output leads or to signal processing circuitry on a monolithic substrate. The electrodes are fabricated out of silicon and contain electrode recording sites of iridium, platinum, or gold. The multiple recording sites along the length of the shank allow for sampling

23

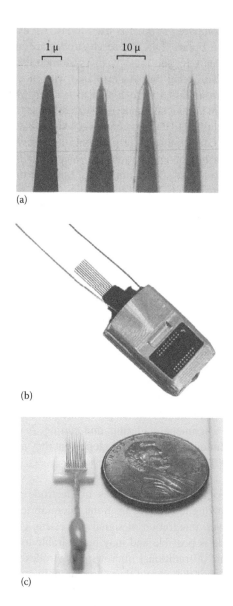

(a)

(b)

(c)

Figure 2.4 Examples of microwire electrodes and arrays: (a) Electromicrograph of sharpened tungsten microelectrodes. (Adapted from Hubel, D. H., Tungsten microelectrode for recording from single units, *Science*, 125(3247), 549–50, 1957.) (b) Tungsten microwire array from Tucker-Davis Technologies. (c) Floating microwire array from MicroProbes. (Salcman, M., and M. J. Bak, Design, fabrication, and in vivo behavior of chronic recording intracortical microelectrodes, *Biomed Eng*, 24(2), 121–8, 1977.)

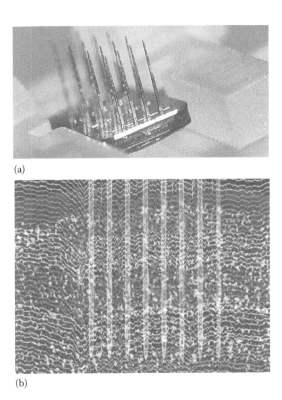

(a)

(b)

Figure 2.5 Examples of planar micromachined electrode arrays. (a) A 3-D array from NeuroNexus. (b) Multiple depth recordings from planar arrays and their corresponding neural signals. (Adapted from Kipke, D. R., W. Shain, G. Buzsáki, E. Fetz, J. M. Henderson, J. F. Hetke, and G. Schalk, Advanced neurotechnologies for chronic neural interfaces: New horizons and clinical opportunities, *J Neurosci,* 28(46), 11830–8, 2008; Buzsáki, G., Large-scale recording of neuronal ensembles, *Nat Neurosci,* 75(5), 446–51, 2004.)

the layered structure of the cortex (Vetter et al., 2004). Figure 2.5b shows multiple depth recordings from planar arrays. Here, the neural recordings are superimposed on the electrode array and tissue to show how the neuronal potentials vary as a function of the depth. Planar micromachined arrays can be fabricated to be "fixed" or "floating" through the selection of what kind of connector or wire is attached to the bulk assembly that is implanted in the brain. In addition, they can be designed to have virtually any length, recording site size, and recording site spatial arrangement. Electrodes typically range from 15 to 50 µm in thickness and can have a variety of widths depending on how many electrodes are placed on the shank. One major consideration for planar micromachined electrodes is that they can be brittle and break either outside or inside the tissues. These electrodes are not approved for human use.

2.3.1.3 2-D bulk micromachined An alternative to the planar microma-chined electrodes is the bulk micromachined electrodes such as the "Utah array." Originally designed for the peripheral nervous system, the Utah array, as shown in Figure 2.6a and b, has become the standard for human neuro-prosthetic applications (Maynard et al., 1997; Hochberg et al., 2006; Collinger

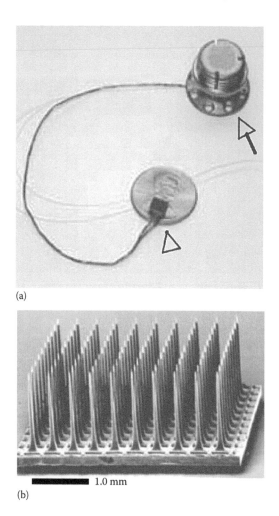

(a)

1.0 mm

(b)

Figure 2.6 Example of 2-D micromachined array. (a) Utah electrode array, flexible cable, and pedestal for connecting back to electronics. (b) Scanning electron micrograph of a Utah electrode array. (Adapted from Hochberg, L. R., M. D. Serruya, G. M. Friehs, J. A. Mukand, M. Saleh, A. H. Caplan, A. Branner, D. Chen, R. D. Penn, and J. P. Donoghue, Neuronal ensemble control of prosthetic devices by a human with tetraplegia, *Nature*, 442, 164–71, 2006.)

et al., 2013). The Utah array is a micromachined square made out of silicon and arrays of 96 single contact electrodes are diced from the bulk substrate as shown in Figure 2.6b. Because the electrode emerges from this substrate, there is a limit to the length of the electrodes. Most electrodes are fabricated to be 1.6 to 1.8 µm to reach layer V neurons of the cortex. The electrodes can either be cut all at the same length or on a slant to account for multiple depth neural targets. The recording sites at the tips contain platinum or platinum/iridium-sharpened recording sites. One advantage of the electrode is that it closely packs a large number of channels in a 4 × 4 mm area. All 96 channels are rapidly injected into the neural tissue using a pneumatic injector that is designed to minimize neural tissue damage during insertion (Rousche and Normann, 1992). The injection can be tricky because the pneumatic piston needs to be directly aligned on top of the array such that as it is pressed into the cortex so that the electrodes uniformly reach the same depth. Misaligned injection leads to arrays that sit on an angle in the cortex. The array is bonded to a demultiplexer connector, which contains a pedestal shown by the arrow in Figure 2.6a. This pedestal is bolted to the skull with titanium screws to create a percutaneous connector to the externalized recording electronics.

2.3.1.4 Electrocorticogram electrodes To record neural activity on the surface of the brain, the electrocorticogram (ECoG) is used and shown in Figure 2.7. The ECoG consists of a sheet of flexible material with electrodes embedded in it. The array just sits on the surface of the cortex either in a subdural or epidural configuration and because of the larger electrode size (compared to indwelling microelectrodes) record mesoscopic neuronal fields (Freeman, 2000). The use of the ECoG array was pioneered by Penfield and Jasper in the 1950s to support their clinical mapping of brain function for the treatment of intractable epilepsy (Jasper and Penfield, 1954). These arrays were very similar to those shown in Figure 2.7a from Ad-Tech Medical Instrument Corporation and consist of a flexible silicone substrate with platinum disks of 4 mm diameter with 1 cm spacing. These electrodes are used widely in medical centers internationally for neurosurgery monitoring and commonly consist of 16, 32, 64, and 128 electrodes. As shown in the figure, these recording disks can be placed in virtually any 2-D geometry to conform to the convexities of the brain surface. The electrodes consist of multiple "tails" that provide connection to the recording electronics. These "tails" are tunneled through notches in the skull bone and traverse the scalp. They are sutured in place and then connected to the electronics as shown in the upper left corner of Figure 2.7a. Although silicone is flexible, the traditional ECoG arrays do not perfectly conform to all of the variations in the gyri of the brain and thus can create a condition in which some electrodes are contacting neural tissue and some are not. Electrodes that do not contact the brain introduce noise. More advanced flexible and foldable electrodes as shown in Figure 2.7b allow for better contact (Kim et al., 2010; Viventi et al., 2011). These electrodes take advantage of the microfabrication

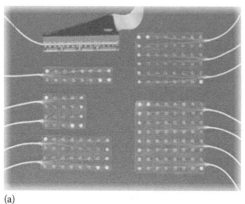

(a)

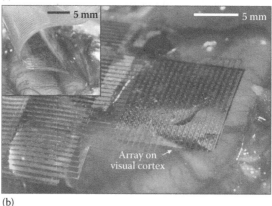

(b)

Figure 2.7 (a) Example of electrocorticogram electrodes. (b) Flexible and foldable ECoG array. (Adapted from Viventi, J., D.-H. Kim, L. Vigeland, E. S. Frechette, J. A. Blanco, Y.-S. Kim, A. E. Avrin et al., Flexible, foldable, actively multiplexed, high-density electrode array for mapping brain activity in vivo, *Nat Neurosci*, 14, 1599–605, 2011.)

techniques described in Sections 3.3.1.2 and 3.3.1.3; however, they are flat and are made of polyimide or similar polymer materials. Unlike the Ad-Tech electrodes, the use of microfabrication allows for virtually any selection of electrode size and spacing. The electrode arrays shown in Figure 2.7b can have hundreds to thousands of electrodes.

2.3.1.5 Depth electrodes To reach deep structures in the human brain, electrodes such as the one shown in Figure 2.8 are used. This model of electrode called the "3387" from Medtronic consists of a quadripolar design (four electrodes). The electrode has a soft blunt tip to help facilitate the movement of the electrode through neural tissue to deep targets. The total length of this electrode is

28

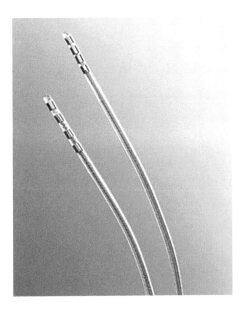

Figure 2.8 Medtronic 3387 deep brain stimulation lead. (Adapted from Liker, M. A., D. S. Won, V. Y. Rao, and S. E. Hua, Deep brain stimulation: An evolving technology, *Proc IEEE*, 1129–41, 2008.)

40 cm with a cylindrical body that has a 1.27-mm diameter. This electrode is most commonly used for stimulating neural tissue and the four 1.5-mm-long electrodes are aligned at the tip. Each platinum electrode also has a spacing of 1.5 mm. The electrode contains a total of four connector wires for each of the contacts and they connect back to the recording or stimulating hardware that is implanted elsewhere in the body. These macro style electrodes do not have the capability to measure the activity of single neurons and either record or stimulate neuronal fields.

Exercises

1. What are two types of microelectrode arrays? Compare and contrast their fabrication, design, and function.
2. You are testing two microelectrodes with exactly the same surface area. However, one electrode has an impedance of 1 MΩ whereas the other has an impedance of 200 kΩ: which electrode would provide the better ability to discriminate action potentials?

   Use the equation $v_n = \sqrt{(4k_B TR\Delta f)}$ to justify your answer. k is the Boltzmann's constant (1.38×10^{-23} J/K), T is the absolute temperature (293 K), R is the resistance, and f is the bandwidth for action potentials.

Chapter 3 Electronics for recording

Distinguishing the signal from the noise requires both scientific knowledge and self-knowledge: the serenity to accept the things we cannot predict, the courage to predict the things we can, and the wisdom to know the difference.

Nate Silver
The Signal and the Noise:
Why So Many Predictions Fail—But Some Don't

Learning objectives

- Specify how electronics should be tailored to capture neural signals.
- Determine the factors that may be contributing noise in a neuroprosthetic system.
- Implement hardware and software techniques to reduce noise and isolate neural signals for neuroprosthetic applications.

3.1 Introduction

Access to the multiscale neural representation for brain–machine interfaces (BMIs) hinges on the ability to chronically sense changes in the parallel and distributed processing capabilities of the brain. The great need to reliably record neural modulation with high signal-to-noise ratios (SNRs) and minimal tissue response (Polikov et al., 2005) has sparked the development of a variety of invasive recording techniques and electrode arrays that are capable of recording the activity of tens to hundreds of neurons (Nicolelis et al., 2003).

From an engineering perspective, to achieve large-scale interfacing with the nervous system, the functional building blocks of a neural implant are the electrodes, the amplification stages, and the transmission of information to a signal-processing unit as shown in Figure 3.1. This outline will serve as a basis of the topics of this chapter. The design constraints are easy to state but much harder

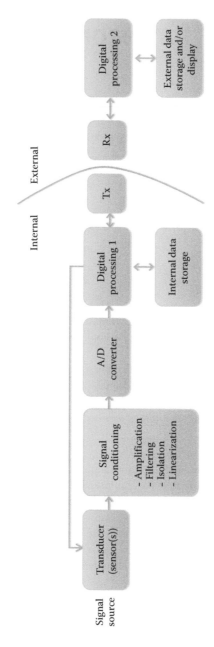

Figure 3.1 Signal-processing block diagram from source to analysis.

to comply with; basically, one wishes to fully integrate the system while minimizing the power consumption per transmitted bandwidth with a given SNR from the electrode to the external receiver. Current chronic electrode technology requires at the surgical stage for the sensor to be placed very close to the neuron for obtaining the highest quality recordings. The electronics of implantable BMIs must meet several challenging design constraints. First, they must be compact and lightweight. Second, they must be low power to allow hours of operation from a small battery supply and to prevent tissue damage due to excessive heating. Implementation difficulties are further complicated by the requirement of wireless operation because wired paradigms are a primary source of infections, discomfort, decreased reliability, and increased acclamation time. Thus, the third constraint is that the implant be powered through the skin, either by recharging batteries or by directly powering the implant. The fourth and arguably the most severe constraint is the small bandwidth (~500 kbit/s) that can be safely transmitted by the wireless link through the skin. As will be seen, the bandwidth is so limited that it is difficult to transmit even a few channels of raw extracellular neural potentials without a compression operation within the implant.

3.2 Use of sensors

The use of sensors is an essential initial first step in the creation of any neural interface system. Without a good sensor that can capture the relevant signal modulation from the brain, no amount of signal processing can be used to remedy the inability to transform thoughts into action (Scott, 2006). By definition, a sensor is a mediator. It converts one or more measured physical quantities into an equivalent signal quantity of another type within a frame of time. Sensors are also considered to be transducers because the convert a physical representation from one form to another. In the nervous system, the problem of transduction is one of converting the physical properties of neuronal activity into machine-readable variables that can be processed. Often in the application of sensors for the brain, a sensor system is used. This system comprises the total signal path from the physical quantity being measured to the observer or hardware acting on the signals and includes all of the conditioning and real-time processing. There are a multitude of subtle differences among sensors that one can define. They include specific hardware such as electrodes, which are electrochemical cells converting charge carriers from ions to electrons or vice versa. An electrode also by design is only a half sensor because two of them are necessary to read the potential difference between two sites in brain tissue. In contrast, a probe is a broader concept than an electrode because it can comprise multiparameter sensors. Probes can come in many forms and can include optical, genetic, or ultrasonic mechanisms of action, all of which could be combined into a single device for measuring signals from the brain (Zhang et al., 2009; Tomer et al., 2014).

3.3 What is a signal?

The signals that are obtained from sensors represent physiological phenomena that convey information about the brain, and they can be described as quantities that cause and reveal the behavior of the organism producing them. The fundamental nature of each signal acquired from the brain prescribes a set of characteristic properties that require appropriate processing methods to reveal the information contained within those signals. The two broadest classification of signals include the continuous (analog) or discrete, as shown in Figure 3.2. Many of the signals of the brain such as local field potentials (LFPs), electroencephalogram, and electrocorticogram are continuous because they involve ionic electrochemical processes that have infinitely fine time resolution; however, those processes can be made discrete and machine readable through a sampling process. To process analog signals by digital means, it is first necessary to convert them into digital form, that is, to convert them to a sequence of numbers having finite precision. This procedure is called analog-to-digital (A/D) conversion, and the corresponding devices are called A/D converters (ADCs). In many cases of interest, it is desirable to be able to reconvert the processed digital signals back into analog form and to apply D/A conversion. The sampling must obey the Nyquist rule, which states that if this signal is sampled at greater than twice the frequency of the highest-frequency component in the signal, then the original signal can be reconstructed exactly from the samples. Continuous signals are defined over a continuous range of a particular variable (usually time), whereas discrete signals are defined at discrete

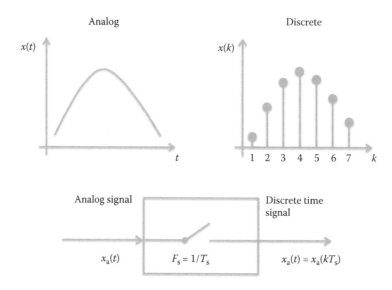

Figure 3.2 Example of continuous and discrete signals.

34

instants (Oppenheim et al., 1999). Here the analog signal is sampled at some rate F_s given by time interval T_s giving rise to a time series of k samples. A/D conversion is achieved in three steps: sampling, quantization, and coding. The process produces a binary-coded number. The digital number represents the input voltage in discrete steps with finite resolution. ADC resolution is specified by the number of bits that represent the digital number. An n-bit ADC has a resolution of 1 part in 2^n. For example, a 12-bit ADC has a resolution of 1 part in $2^{12} = 4096$. A 12-bit ADC resolution corresponds to 2.44 mV for a 10-V range.

For the recording and monitoring of electrical signals in the brain electrodes, amplifiers, filters, ADCs, displays, and a computer processor are needed. On the front end, the electrodes transduce the neurophysiologic signal such that amplifiers can bring miniscule physical phenomena up to a larger level. In other words, amplifiers increase a low input signal into the higher output signal. In neuroprosthetic applications, differential amplifiers are commonly used to help eliminate common-mode noise in signals. They typically consist of two amplifiers with a common output clamp. Both amplifiers have the same degree of amplification, but one of them serves as inverting such that the two output signals can be subtracted. Any common noise elements on both signals will be eliminated. Basic parameters for the design of neuroprosthetic amplifiers include specifying the discriminating factor of the amplifier input (output) specifications, gain, and width of transmission (frequency) band. Filters are needed for this transmission band to remove noise or select signals of interest. Once preprocessed, signals are ready for conversion into the digital domain, which a computer can process and store the resulting time series. Computers can record the amplitude modulation and store the parameters of a signal within memory. This process can be initially online but also switch to offline mode for subsequent processing and evaluation. An example set of signals relevant for neuroprosthetics is provided in Table 3.1. Relative amplitude and frequency ranges are important for these signals because they can often contaminate each other. All signals have a bandwidth or difference between the upper cutoff frequency f_2 and the lower cutoff frequency f_1. Front-end

Table 3.1 Relative Amplitude and Frequency Range for Common Neural and Nonneural Physiological Signals

Signal	Name	Amplitude (mV)	Frequency Range (Hz)
ECG	Electrocardiogram	0.5–5.0	0.01–250
EEG	Electroencephalogram	0.01–50.0	0.1–150
EMG	Electromyogram (surface electrode)	0.1–10.0	0.01–10000
EMG	Electromyogram (needle electrode)	0.05–5	0.01–10000
LFP	Local field potential	0.01–1.0	0.1–300
SUA	Extracellular neural recording	0.05–0.2	300–8000

sensors and amplifiers should be tailored to these characteristics to best capture these signals with the highest fidelity. For the recording of nonelectric signals, all of the hardware is the same, except for the front end, which would replace an electrode with a transducer. The transducer may have specialized properties needed for processing the resulting signals. For example, the front-end hardware needed for an all-optical transducer would require special considerations in terms of power, wavelength, and sensitivity.

In the next sections, a deeper analysis of each of the amplification, filtering, and signal preprocessing steps will be provided. The engineering design of the components will be framed in the context of a neuroprosthetic system because the requirements are unique when compared against general purpose data acquisition devices.

3.4 What is noise?

The input signal to a neuroprosthetic amplifier includes the desired neuropotentials, the undesired neuropotentials, the power line interference and its harmonics (60 Hz in the USA and 50 Hz in Europe), the interference signals generated by the tissue–electrode interface, and the noise from any other source in the vicinity of the sensor. All other signals outside of the desired neuropotentials can be considered to be noise, and minimizing the degradation of the desired signal by noise is of main concern. A common measure of the noise amplitude is the square root of the time-averaged mean. For a given waveform $x(t)$, the root-mean-square (rms) value is defined as $X_{rms} = \left(\frac{1}{T} \sum_0^T x(t)^2 dt \right)^{1/2}$. The quantification of noise and SNR is a science in itself; the experimentalist primarily needs methods for recognizing the types of noise often encountered during specific recording and for minimizing their amplitude and effect. The identification of the noise source will specify the noise elimination methods that will be effective and feasible. These include software or hardware filtering, averaging to reduce the effect of uncommon signal sources, and blanking, which clamps signal acquisition to zero during known periods of high noise.

3.5 Biopotential amplifiers

For electrical interfaces to the brain, the first stage of processing of single neuron recordings is to amplify the extracellular neural signals from the electrodes, which have peak-to-peak amplitudes of 50–500 µV. Amplifiers for the brain have several basic requirements, which include that the measured signal should not be distorted, the physiological process to be monitored should not be influenced

36

in any way by the amplifier, and the amplifier should provide the best possible separation of signal and interference. In addition, these signals suffer large direct current (DC) offsets across different electrodes due to electrochemical effects at the electrode–tissue interface. The magnitude of these DC offsets is 1 to 2 V, much larger than the neural signals to be measured. As described in Table 3.1, the relevant frequencies of the extracellular action potentials range from 300 Hz to 8 kHz, whereas the LFPs extend to less than 1 Hz. Thus, the ideal band-pass filter must reject the DC offset while passing the LFP signal.

Several groups have designed low-noise, low-power amplifiers suitable for neuro-prosthetic implants (Wise et al., 2004), and these custom amplifiers consist of a low-noise 40-dB preamplifier fabricated in the 0.6-µm AMI CMOS process (Harrison and Charles, 2003). To achieve a low cutoff frequency, a large resistance is needed. Because noise performance is important in neural preamplifier designs, noise can be minimized by carefully choosing the width and length of the transistors. However, there is a trade-off between stability and low noise. Because low noise is critical in neuroprosthetic applications, decreasing the transconductance can minimize the thermal noise. Flicker noise, which is important in low-frequency applications, is minimized by increasing the device area.

For the design of amplifier systems for neuroprosthetics, the preamplifier as shown in Figure 3.3 represents the most critical part of the architecture because it sets the stage for the quality of the biosignal. When properly designed, the preamp can minimize most of the signals interfering with the measurement of neural signals. The total source resistance, including the resistance of the biological source and all transition resistances between signal source and amplifier input, causes thermal voltage noise with an rms value of $E_{rms} = \sqrt{4kTRB}$, where k is the Boltzmann constant, T is the absolute temperature, R is the resistance in ohms, and B is the bandwidth in hertz. High SNRs thus require specialized amplifiers with limited bandwidth. The exact relationship between noise and frequency

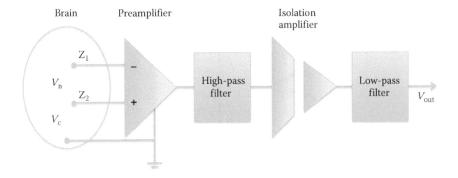

Figure 3.3 Block diagram of a biopotential amplifier.

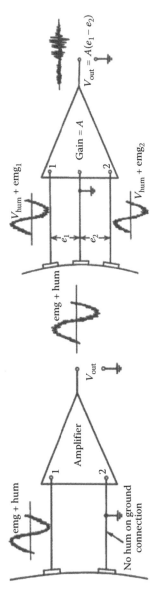

Figure 3.4 A comparison of single-ended and differential amplifiers. (From Winter, D. A. *Biomechanics and Motor Control of Human Movement.* Hoboken, NJ: John Wiley & Sons, 2009.)

depends on the technology, such as the kind of transistors used at the amplifier input stage. Continuing from left to right in Figure 3.3, the purpose of the high-pass and low-pass filters is to eliminate half-cell interference potentials as well as provide a lower and upper bound on the amplifier bandwidth. The isolation stage serves to decouple the brain from the recording equipment and to eliminate safety issues from being directly connected to a power source. Most isolation stages are optical interfaces, which also prevent ground loops. Digital isolation amplifiers use a voltage/frequency converter to digitize the signal before it is transmitted.

The type of amplifier depicted in Figure 3.3 is a differential amplifier. As shown, three electrodes are used. The first two collecting the neural signal V_n, and the third provides the reference potential V_c. The output is the difference between the two input signals scaled by a gain factor. Figure 3.4 compared the differences between a differential amplifier and a single-ended amplifier. In a single-ended design, any noise or hum that appears on the biopotential sensor will also appear in the output because it is referred directly to ground. There are no common sources to eliminate. In contrast, in a differential amplifier, the desired signal appears as a voltage between the two input terminals (input and reference) of the differential amplifier, but it is also compared to an additional signal called the common-mode signal, which appears between inputs and ground. Any common signal applied to both the input and reference should result in an elimination of this common signal at the voltage output.

The common-mode rejection ratio (CMRR) is the ratio of the amplifier's differential gain to the common-mode gain and is a measure of how closely the amplifier approaches an ideal differential amplifier. In this formulation, the common-mode gain is defined as the ratio of the output voltage to the common input voltage. In most cases, the common-mode gain is much less than one. The differential gain is defined as the ratio of the output voltage to the applied differential input voltage, and it is usually much larger than one. The CMRR is expressed in decibels (dB) and should be on the order of at least 100 dB. As shown in Figure 3.3, the rejection of the common mode is a function of the source impedances Z_1 and Z_2. In an ideal case if $Z_1 = Z_2$, the output voltage is the pure neural signal amplified by the differential gain of the amplifier. When Z_1 and Z_2 are not equal, the common-mode signal causes currents to flow through Z_1 and Z_2, and the related voltage is amplified and not rejected by the differential amplifier.

3.6 Filtering

Filtering is a neural signal-processing operation that alters the frequency content of a signal and allows for the specification of certain modulations of neural activity to be used in other components of the interface systems. Examination of Table 3.1 reveals that neural activity exists in bands. For example, single unit activity (SUA) can be identified at frequencies higher than 300 Hz whereas LFP exists at frequencies less than 300 Hz. To obtain these specific activities, filters must be designed such that they are frequency selective in which they pass one band of frequencies and reject (or attenuate) signals in other bands. In this way, filtering can also be used to enhance the SNR or eliminate certain types of noise, provided that the system designer knows at which frequencies the noise appears.

3.6.1 Filter design

The frequency response defines a filter's behavior. To properly design a filter tailored for neuroprosthetic applications, one must specify the gain and phase information through mathematical functions that have real and imaginary components (i.e., complex functions). Because complex functions are related to oscillatory signals, the gain and phase responses specify how the filter alters the amplitude and phase of sinusoidal inputs produce modified sinusoidal outputs. The gain and phase responses show how the filter alters the amplitude and phase of a sinusoidal input signal to produce a sinusoidal output signal. The modifications to the signals that are input and output from the filter are often expressed through the dimensionless decibel (dB) scale, which relates either two signal powers or amplitudes and takes the following form: $10\log_{10}\left(\dfrac{P_1}{P_2}\right) = 10\log_{10}\left(\dfrac{A_1}{A_2}\right)^2 = 20\log_{10}\left(\dfrac{A_1}{A_2}\right)$.

Designers of neuroprosthetic systems that contain filtering operations must have extensive electrical engineering knowledge of the intricacies of how to design the best performing filters to address the signal-processing needs. An overview of the critical aspects of filtering are presented here as a foundation for neuroprosthetic applications, and they include specification of the properties of the filter such as which frequencies are attenuated and which are not, selection of an architecture to conduct the filtering operation, and understanding what assumptions that architecture imposes on the signals themselves.

3.6.1.1 Pass band and stop band   When designing a neuroprosthetic filter, the first consideration is to determine the frequencies that will be used to define the pass band and the stop band. The pass band is the range of frequencies in which there is little attenuation, and the stop band is the range of frequencies where the signal is attenuated. Using Table 3.1 to identify SUA, one would design a filter with a pass band of 300 Hz to 6 kHz and stop bands for all other frequencies. In this example, there are two boundaries between the pass band and the stop band, one less than 300 Hz and one more than 6 kHz. Figure 3.5 shows an example filter and its corresponding pass and stop bands. Depending on the filter's mathematical formulation, the pass bands cannot be perfectly flat and contain some amount of ripple or variation in the gain as a function of frequency (Oppenheim et al.,

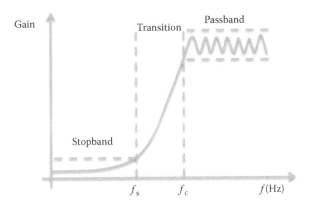

Figure 3.5 Filter characteristics.

1999). These ripples add distortion and should be minimized. Most filter design tools allow experimenters to specify how much ripple can be tolerated for a given signal. However, this can come at the cost of filter order.

3.6.1.2 Filter order The order of the filter relates signal attenuation and increases by a factor of 2 for each doubling of the frequency (octave). The filter order can introduce a second distortion called the group delay, which produces time delays in certain amplitude envelopes of the time-domain signal. Group delay is particularly important for neuroprosthetic applications because filtered neural signals with larger group delays can cause desynchronization of those signals with respect to other physiological signals, which are not filtered. It is important to carefully choose the filter order so that such time distortions are minimized or perform correction of the group delay through forward-reverse filtering. However, the latter technique may not be applicable for real-time testing. A third component of distortion can also come from the phase shift frequency. Distortion is minimized if the phase shift in the pass band is linearly dependent on the frequency of the sinusoidal components. The last major component of the filter design is the frequency where the signal response starts to be attenuated. It is called the cutoff frequency (f_c). This begins the transition band, and there is a characteristic frequency called f-3 or the −3 dB frequency within it, where the amplitude of the input signal falls to $\frac{1}{\sqrt{2}}$. The sharpness or how steep the roll-off is designed on a filter also affects filter order and should also be carefully considered given the application.

3.6.1.3 Finite impulse response and infinite impulse response filters To achieve the desired filtering specifications such as pass band, stop band, and ripple, one must also choose a filter topology to implement the filter design. The topology

41

refers to the layout of connections between input and output and the mathematical operations performed on those signals. Two classes of filters are commonly used: finite impulse response (FIR) and infinite impulse response (IIR). FIR filters are nonrecursive filters because the output depends only on a weighted sum of the input signals. IIR filters are known as recursive filters because they use feedback by delaying the output and rerouting it to the input. For an equivalent FIR design, the IIR often results in a lower filter order because it uses feedback, which represents previous versions of the signal and is more efficient in the use of time information. One major drawback from IIR filters is that they do not exhibit linear phase and can introduce distortions, as described earlier. They also have the quality that they can be unstable because they positively reinforce themselves through the use of feedback. FIR filters are linear phase constraint and have the advantage of not altering the phase of the signal.

3.6.1.4 Hardware versus digital filters   Depending on where it is implemented in neuroprosthetic applications, filtering can either be performed directly in either hardware or software as a digital filter. If the filtering is implemented in hardware, real electrical circuits consisting of capacitors, resistors, inductors, and operational amplifiers perform the corresponding filter specifications. Hardware filters are not general purpose in the sense that they are designed for one set of filter specifications and require a specific set of components to achieve those specifications. If a new filter is required, a new filter hardware design is needed. When a specific known filter with efficient processing and size constraints is needed, hardware filters are unmatched compared to alternatives. The more general approach to filtering is to use a digital computer that can be programmed to implement a filter algorithm in software. This approach has the advantage of flexibility, where the designer can choose a variety of filter architectures, band passes, ripple constraints, and filter orders to assess the filter performance and trade-offs for each application. The trade-off is that these digital filters implemented mathematically can be slower and less efficient than fit-for-purpose hardware designs.

3.7 Adaptive filters

The design of the filters in Section 3.6 assumes that the signal and the noise are stationary, which means that their statistics are constant over time. Neuroprosthetic applications are almost exclusively defined by nonstationary signals. Neural signals are constantly changing as a function of the environment, learning, and affective state. When used during activities of daily living, neuroprosthetic systems can also be affected by noise from a variety of sources, each of which has their own characteristic properties. For these types of nonstationary signals, it is difficult to specify the filter a priori because any initial filter designs may be ineffective for changes in the filter frequency band and signal magnitude as time progresses. For these scenarios, adaptive optimal filters should be used and are well designed for applications of noise cancellation. This class of filters has the special property in which they can adjust their own parameters based on signal characteristics (Haykin, 1996). At the core of an adaptive filter is an adaptation algorithm that specifies the mathematical operations that monitor the input, output, and environmental signals and varies the filter's transfer function. On the basis of the actual signals received, the adaptive system attempts to find the optimum filter design. Figure 3.6 shows the architecture or signal flow graph for an adaptive filter. In this example, one seeks to automatically remove noise from a contaminated neural recording. For adaptation to be possible, two signals are needed: one sensor with the noisy neural signals and another sensor with a reference noise source from the environment. Adaptive filtering first involves the filtering process where an input signal $u(t)$ is passed through the filter to produce an output $y(t)$. The structure of the filter can be either FIR or IIR and is designed based on the characteristics of input signals. No other frequency response information or specification information is provided. The blow-up box in Figure 3.6 shows the signal flow for a single input FIR filter. Here, the input is fed through a tapped delay line and then scaled by the filter weights (W). The weighted sum becomes the adaptive filter output $y(t)$. Once an output is produced, the adaptation process can begin where the filter

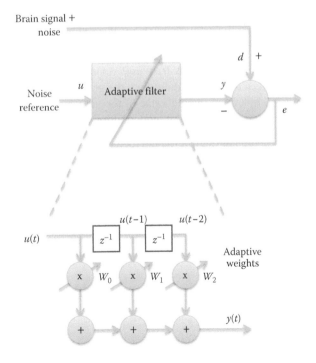

Figure 3.6 Components of an adaptive filter.

43

coefficient W can be adjusted. A common optimization criterion for the adaptive filter is to minimize the mean square error (MSE) between the desired signal $d(t)$ and the filter output $y(t)$. Additional information about this optimization process will be provided next.

The least mean square algorithm is one of the most popular optimization approaches for adaptive filters. The cost function to be optimized is defined as the MSE. The MSE corresponds to the expected value of the squared error loss or quadratic loss. Because it is the second moment, the MSE is powerful because it contains both the variance and the bias of the estimator. The optimization process is based on the method of steepest descent. In the space of the filter parameters or weights, the MSE forms what is called an error surface. Certain values of the weights create larger errors between the filter output and the desired signal, whereas other values create smaller errors. The most optimal filter is the one that adapts itself in the direction of the smallest error. Descending on the error surface to this minimum in the steepest manner possible is desirable. Mathematically, the steepest descent is defined by computing the gradient or partial derivative of the MSE with respect to the filter weights. In the example given in Figure 3.6, the estimation error is $e(t) = d(t) - y(t)$, and MSE $= \sum [d(t) - y(t)]^2$. Taking the partial derivative of the MSE with respect to W invokes the chain rule and yields $2u(t)e(t)$. The input $u(t)$ appears in the partial derivative because the derivative of $y(t)$, which is included in the error, depends on W, as shown by the filter output $y(t) = \sum_{k=0}^{M-1} u(t-k)w_k^*(t)$. Putting all of the mathematics together, the tap-weight adaptation is given by $W_k(t + 1) = W_k(t) + \mu u(t - k)e(t)$. To get to the error surface minimum, at every iteration of the time t, the gradient is estimated, and the weights are updated by taking the old value of the tap weight and adding to it a scaled version (learning rate mu) of the input multiplied by the error.

The method is very computationally efficient for electronics that are used in neuroprosthetic applications, which require low power and low complexity. The update only consists of a few multiplies and an addition. For a naïve system, the adaptation begins with seeding the weight values with small random numbers between −1 and 1. An input is fed to the system, and an output is produced. The MSE and gradient are computed, and then the weights are updated. Time advances, and another input is fed to the system. MSE, gradient, and weight update again follow as before. The process continues until an acceptable MSE for the application can be achieved.

3.8 Conclusion

This chapter began with defining constraints for the design of electronics for recording and processing signals in neuroprosthetic applications. Neural signals have unique properties in terms of their signal frequency attributes. If the front-end sensors and back-end electronics are not properly designed, the signals acquired from the neural circuit may not be accurately represented as distorted, which will impede the primary goal of neuroprosthetics to translate thoughts into action. Because neuroprosthetics are designed to be deployed in awake and behaving applications, the electronics need to accurately be able to identify what is signal and what is noise. Traditional electrical engineering approaches to hardware or software filters can produce signal conditioning options, but only in limited applications where the signal properties can be specified a priori. Neural signals often contain many nonstationarities, which require advanced techniques such as adaptive filters, to continually optimize themselves to track time-varying properties in the signals. Ultimately, the neuroprosthetic system that produces the most high-quality and high-resolution signals is the one that is most desirable. With uncontaminated accurate signals, the process of neural decoding is facilitated by producing a "window" into the processing capabilities of networks of the brain.

Exercises

1. You seek to engineer a data storage system for a neural signal recorder. How big of a disk will you need for recording 1 hour of data if the system has a 12-bit ADC, 100 electrodes, and a sampling rate of 24 kHz?

   You would also like to transmit these data wirelessly to a base station for storage. How much bandwidth would you need to design into your wireless?

2. Consider that you are performing a neural recording (EEG or microelectrode) in two subjects. During your experiments with both subjects, you find that your recordings are contaminated with noise. In subject 1, you determine that the noise is coming from the environment and is affecting all of the electrodes. In subject 2, you determine that the noise is coming from the muscles related to eye blinks and only affecting a subset of your electrodes.

   Describe two ways that you would engineer your referencing system for minimizing the noise contamination in these two scenarios. What are the strengths and weaknesses of each approach?

Chapter 4 Surgical techniques for implantation and explantation of microelectrode arrays

"Why do you think it is…" I asked Dr. Cook… "that brain surgery, above all else—even rocket science—gets singled out as the most challenging of human feats, the one demanding the utmost of human intelligence?" [Dr. Cook answered] "No margin for error."

Michael J. Fox
Lucky Man (2002, 208)

Learning objectives

- Understand the level of precision and amount of detail needed for a neuroprosthetic implant surgery.
- Plan and conduct the steps needed to perform stereotaxic implantation of neuroprosthetic electrodes.
- Validate electrode placement.

4.1 Introduction

The process of implanting electrodes into the brain to capture the coordinated activity of neural populations is the single most important step in the design of neuroprosthetics. There is an old wives' tale that says, "One cannot bake good cakes with rotten eggs." The same holds true in the design of neuroprosthetics. If the signals that form the basis of control of prosthetic devices are not of the highest quality or if they do not contain the relevant functional neuromodulation related to the task at hand, no amount of engineering or system design can be used to remedy the control problem. With this being the case, the surgical process of chronically implanting electrodes into the brain must be approached with the utmost care, planning, and exquisite technique to ensure the most beneficial outcome for the subject and to increase the chances of obtaining the best signals. The surgical approach for successful, long-term implantation of electrode arrays should be consistent with the following concepts: maximal precision, minimal invasiveness, maximal choreography, minimal intraoperative time, and minimal trauma to the brain. Both quantitative and anecdotal evidence have shown that invasive, long procedures that introduce trauma to the brain will reduce the amount of accessible neural activity and affect beneficial, long-term outcomes of the implant procedure. The purpose of this chapter is to provide a comprehensive didactic and procedural guide to the implantation and explantation of microelectrode arrays.

4.2 Targeting

The first step in any implant procedure is the determination of the insertion sites and final targets for the electrode arrays. These sites are prescribed by the need to acquire systems-based neurophysiological signals from the sensory, motor, limbic, or associative neural structures in the cortical or subcortical regions of the brain. Detailed planning involves the consideration of the trajectory of electrode insertion to maximize hitting the desired target and the necessary precautions that are to be taken to avoid intersecting vasculature and/or the ventricular system. Damaging or disrupting these structures can lead to hemorrhage, stroke, and recovery complications in the subject and are to be avoided. In addition to the trajectory of any single array, the interactions of multiple arrays must be considered as part of the presurgical plan. This includes their spatial layout to conform to the topographic organization of neural structures as well as the ability to physically fit all of the hardware into the available intracranial or extracranial area. The locations of access into the skull (through craniotomies or bone windows) need to be balanced against the final location of electrode connectors, anchoring screws, grounding/lead wires, and preamplification hardware (headstages). Presurgical targeting should be conducted with the aid of a stereotaxic atlas and MRI if it is available. A variety of atlases are available for common animal models used in the neuroprosthesis literature and a list is provided below. The process of targeting is completed when a list of coordinates is determined for each target. To illustrate this concept, an example of a complete targeting procedure is presented next.

1. Rat
    a. Paxinos G., Watson C. *The Rat Brain in Stereotaxic Coordinates*. Fourth Edition. New York, Elsevier, 2004. With CD-ROM.
    b. Paxinos, G., Ashwell, K. W. S., Tork, I. *Atlas of the Developing Rat Nervous System*. Second Edition. New York, Academic Press, 1994.

2. Rhesus monkey
   a. Paxinos, G., Huang, Xu-Feng, Petrides, M., Yoga, A. *The Rhesus Monkey Brain in Stereotaxic Coordinates*. Second Edition. New York: Elsevier, 2008.
   b. Saleem, K., Logothetis, N. A. *Combined MRI and Histology Atlas of the Rhesus Monkey Brain in Stereotaxic Coordinates*. Second Edition, New York: Academic Press, 2012.
3. Marmoset monkey
   a. Palazzi, X., Bordier, N. *The Marmoset Brain in Stereotaxic Coordinates*. New York: Springer: 1 SPI Edition, 2008.
   b. Paxinos, G., Watson, C., Petrides, M., Rosa, M., Tokuno, H. *The Marmoset Brain in Stereotaxic Coordinates*. New York: Elsevier, 2011.
4. Guinea pig
   a. Luparello T. J. *Stereotaxic Atlas of the Forebrain of the Guinea Pig*. Baltimore, MD: Williams & Wilkins, 1967. Text in English, German, French and Spanish.
   b. Shao, D. H., Qian, X. B. Stereotaxic localization of the brain stem auditory centers in guinea pigs (*Cavia porcellus*). *Sheng. Li Hsueh Pao* 40(4): 386–389, 1988.
5. Human
   a. Schaltenbrand, G., Wahren, W. *Atlas of Sterotaxy of the Human Brain*. Stuggart: Georg Thieme Publishers, 1977.

4.2.1 Targeting example

In this example, a step-by-step procedure is provided for targeting both a cortical and subcortical neural structure. After working through this example, one should be able to select any neural target, identify its coordinates on a stereotaxic atlas, indicate relative areas to avoid or pass through, develop an insertion procedure, and assemble a table outlining the surgical targeting plan. These components to electrode placement are presented in the targeting of the motor cortex and ventral striatum shown below. The animal model used here is the marmoset monkey.

4.2.1.1 Hand and arm region of the primary motor cortex   The first target presented here is the hand and arm region of the primary motor cortex. This target is necessary for the acquisition of motor intent related to volitional control of the upper extremity. In most neuroprosthetic applications, localization of layer V neurons is desirable because they form cortical–spinal connections during the expression of intent. Upon determination of the primary motor cortex as the target, the next step is to consult a detailed physiological functional mapping study of the prescribed area. These are commonly found in the literature and often involve electrical, optogenetic, and cytoarchitectural techniques to identify

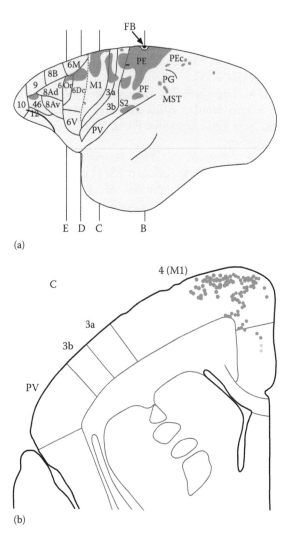

(a)

(b)

Figure 4.1 Functional localization motor neural representation using electrical stimulation. (a) Identification of the motor strip as indicated by pointer "C." (b) Laminar organization of neurons at the coronal slice indicated by "C" that produce motor activation to electrical stimuli. These indicate the depth at which to target the electrode positioning. (*Continued*)

(c)

Figure 4.1 (Continued) Functional localization motor neural representation using electrical stimulation. (c) Cross section of the marmoset five-layer cortex in M1. Layer V neurons are located approximately 1.6 to 2.1 mm from the cortex (scale bar: 200 µm). (Reprinted with permission from Burman, K. J., S. M. Palmer, M. Gamberini, M. W. Spitzer, and M. G. P. Rosa, Anatomical and physiological definition of the motor cortex of the marmoset monkey. *J Comp Neurol* 506 (5), 860–76, 2008, doi: 10.1002/cne.21580.)

regions of interest and their related functions. A representative marmoset monkey brain is shown in Figure 4.1a, and it delineates the major divisions of the frontal cortex. In addition, the shaded areas indicate locations where electrical stimulation was applied to evoke functional responses. In this example, the target is indicated by M1 and the cortical surface runs medial to lateral as indicated by pointer "C." Typically in higher-order primates and humans, the general layout of M1 is such that the face representation is mostly lateral and working medial, followed by the forelimb, trunk, and hind limb, respectively. Approaching the lateral sulcus, the cortical topography merges back into PMd and/or the somatosensory cortex. Thus, it is advantageous to stay at least a couple of millimeters away from the sulcus. The relative shape of the target area and the surrounding structures in an important consideration because the electrode array that is used to capture neuronal activity should be custom-tailored to fit into the target. For the M1, the largest coverage of the cortical surface would be acquired using a rectangular-shaped array or set of arrays that are oriented so that their long axis runs medial to lateral as in the animals' anatomical motor cortex. It is also important to note that the M1 does not run precisely medial to lateral but has a slight curvature to the rostral part of the brain. This additional detail will be used later to properly orient the electrode array to be implanted. In Figure 4.1b and c, the cross-sections of the cortex are presented. Layer V neurons (indicated by oval) are located approximately 1.6 to 2.1 mm from the cortex. This measurement is used to guide the depth of the electrode drive during surgery. All cortical and subcortical targets should be confirmed with electrophysiological recordings. Changes in the properties of the electrophysiological recordings can be used to map the layers of the cortex. Typically, as electrodes approach layer V of the cortex, single neural action potentials increase in amplitude. In addition, as the electrode passes from above to below the axon hillock, the polarity of action potentials can flip from positive to negative. Utilizing the properties of the dipole formed by the bulk of neurons in layer V of the cortex is valuable in building confidence in electrode placement (Buzsáki, 2004). Combining stereotaxic coordinates, anatomical structures, and electrophysiological recordings maximizes the information needed for accurately and consistently placing electrode arrays into the brain.

Once the relative size and shape of the M1 target area has been determined, the next step is to determine its stereotaxic coordinates. Most atlases use either the bregma or the interaural line as the zero point from which to make all measurements. The bregma is defined as the point of intersection of the sagittal skull suture with the coronal suture. Often, a curve of best fit is needed to define the intersection due to variability in each animal's sutures. In contrast, the interaural line is defined as the line connecting the tips of the ear bars of the stereotaxic instrument and the top of the brain. Both bregma and interaural line are viable choices for the zero point if the use of each depends on the stereotaxic atlas that is being referenced. For the examples presented here, the bregma is selected as the zero point. In Figure 4.2a, the top of the marmoset skull is shown in a stereotaxic frame. The two pins forming a line passing through the bregma and perpendicular to the sagittal suture show the identification of the bregma. Once the bregma has been identified, the next step is to consult a stereotaxic atlas and find the location of the primary motor cortex. In this example, the atlas titled *The Marmoset Brain in Stereotaxic Coordinates* was consulted and the motor cortex was indicated to span the coordinates: 2.42 to 6.2 mm anterior–posterior (AP) from the bregma and 2.7 to 5 mm medial–lateral (ML) from the midline (Palazzi and Bordier, 2008). Figure 4.2b shows a coronal slice of the brain at 5.50 mm anterior to the bregma (as indicated by the + symbol). This slice is positioned on the rostral end of the primary motor cortex. In the top right corner of Figure 4.2b, a cartoon drawing of the brain is shown with a vertical line indicating the relative position of the coronal slice. Comparing this location with respect to the lateral sulcus and the location of the motor strip in Figure 4.1a, there is good agreement between the functional location of the primary motor cortex and the stereotaxic target. Consolidating all of the information together, two coordinates need to be determined: (1) the location of the skull craniotomy and (2) the location of the brain target itself. For this example, the final location of M1 craniotomy relative to the bregma would be AP 3 to 6 mm and ML 2 to 6 mm. The center of the electrode array insertion site (before onsite surface stimulation)

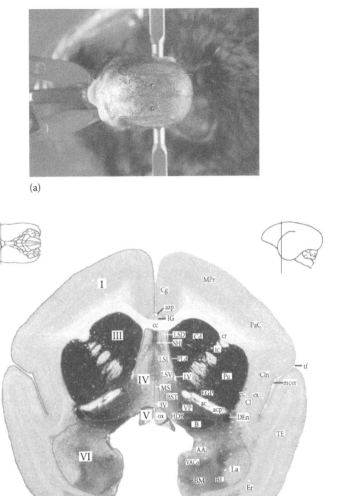

(a)

(b)

Figure 4.2 Targeting the hand/arm region of the primary motor cortex.
(a) Localizing the intersection of the sagittal and coronal sutures to define
bregma. (b) Stereotaxic location of the hand and arm region of M1. (Reprinted
with permission from Palazzi, X. and N. Bordier, *The Marmoset Brain in
Stereotaxic Coordinates*, Springer Verlag, 2008.)

53

would be located at 4.5 mm AP and 4.5 mm ML. To target the layer V neurons, the electrodes would need to be at least 5 mm long to reach a depth of 2 mm (and to take into consideration the cranial/skull thickness of 1–2.5 mm). Given the geometry and layout of M1, all of the electrodes in the array would need to fit into a 12- to 16-mm^2 area to cover the forelimb/hand region of M1.

4.2.1.2 Nucleus accumbens and ventral striatum   The procedure for implanting electrodes into deep brain structures is very similar to cortical structures. However, special consideration must be taken to avoid vasculature, ventricles, or lesioning of brain structures that are critical for behavioral or autonomic function. The best approach for avoiding damage during deep brain implantation is to first visualize in three dimensions the subcortical targets, their related structures, and associated major support systems (vasculature and ventricles). To illustrate this point, Figure 4.3a shows a 3-D rendering of the morphology of the striatum. The striatum has a complex shape, in which it tapers moving rostrally to caudally in the brain. It also contains curvature moving dorsally to ventrally. Such organic and variable structures are quite different from the design of electrode arrays that are often fabricated with very generic rectangular or cube geometric designs. For this example, the target is the nucleus accumbens, which is indicated by a star in Figure 4.3b (Pennartz et al., 1994). When targeting deep brain structures, it is often desirable to custom-design the electrode array to conform to the desired target geometry. This can be achieved through custom arrangement of the wire positioning in the array as well as by modifying the lengths of multiple electrodes in the array to conform to the varying depth of the structure. In the case of the nucleus accumbens, the target is somewhat "football"-shaped with the long axis running anterior to posterior. For this structure, its largest axis is approximately 2 mm; therefore, electrodes should be arranged into a more rectangular shape. Once the custom geometry of the electrode has been determined, the next step is to develop a surgical plan for the insertion trajectory of the electrode array into the deep brain structure. In Figure 4.3b, there are multiple considerations to account for before implanting the array. First, notice the large ventricles highlighted in red that are near the midline. In addition to the ventricles, there is often vasculature that accompanies these structures as they move throughout the brain. In all circumstances, it is undesirable to drive electrodes near or through these areas because they can cause serious complications related to cerebral spinal fluid leakage and or stroke. Therefore, a surgical plan for driving the electrode array directly vertical from the cortex is not an option. As an alternative, the electrode should be driven from an oblique angle as shown by the green triangle in Figure 4.3b. In this case, there are multiple angles from the vertical that can be chosen; the larger the oblique angle, the larger the safety margin for hitting the ventricles. There is one trade-off that as one moves more lateral, it becomes more difficult to open a craniotomy due to the musculature located on the sides of the skull. As a result, the final location of the craniotomy and angle of insertion are a compromise among benefits to the subject, ease of surgical implantation,

Lateral view Superior view Posterior view

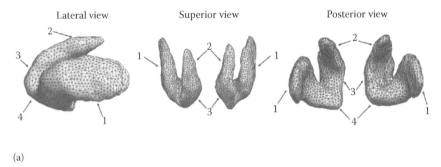

(a)

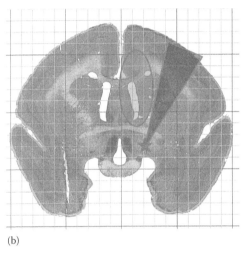

(b)

Figure 4.3 Targeting the complex interaction among deep brain structures. (a) 3-D rendering of the striatum. Note the tapering shape of the caudate. Such variations should be accounted for when matching electrodes to targets for chronic microelectrode array implantation. 1. Putamen; 2. Body of caudate; 3. Head of caudate; and 4. Nucleus accumbens. (Reprinted with permission from Koikkalainen, J., J. Hirvonen, M. Nyman, J. Lötjönen, J. Hietala, and U. Ruotsalainen, Shape variability of the human striatum—Effects of age and gender. *Neuroimage* 34 (1), 85–93, 2007, doi:10.1016/j .neuroimage.2006.08.039.) (b) Trajectory considerations when targeting deep brain structures. Range (dark gray) of low-risk trajectories for implantation of microelectrode arrays for targeting the ventral striatum (nucleus accumbens). Regions to avoid include ventricles and related vasculature (light gray). Multiple angles of insertion can be chosen depending on the target and available room on the skull. (Reprinted with permission from Tokuno, H., I. Tanaka, Y. Umitsu, and Y. Nakamura, Stereo Navi 2.0: Software for stereotaxic surgery of the common marmoset [*Callithrix jacchus*]. *Neurosci Res*, 65 (3), 312–15, 2009, doi: papers2://publication/doi/10.1016/j.neures.2009.08.004.)

and ability to fit all of the necessary hardware. In this example, one possible set of coordinates for targeting the nucleus accumbens includes placing a craniotomy 3.5 to 8.5 mm (AP) and 2 to 5 mm (ML) relative to the bregma. For the insertion site, the stereotaxic drive arm could be rotated to an angle of $30°$. This aims for the lateral portion of the nucleus accumbens with a deep brain target of 6.0 mm (AP) and 6.0 mm (ML). For a rectangular array that spans 2 mm anterior and 2 mm posterior to the target, this would result in a coverage of 4.0 to 8.0 mm AP. Given that this is a deep brain structure at a depth of 11 mm, the recommended electrode length would be 14 mm to incorporate a skull thickness of 1 and 2 mm for the gap between the bottom of the skull and the top of the brain.

To find the appropriate set of benefits and risks for all of these options, there are multiple software packages that can be used for targeting in humans and in animals. In the example shown here, StereoNav was used to identify the stereotaxic targets, to analyze relative brain structures anterior and posterior to the target, and to visualize cellular lamina and fiber tracks (Tokuno et al., 2009). Visualization of these ancillary targets is important for also increasing the accuracy of the targeting. For example, in this trajectory for implanting electrodes into the nucleus accumbens, there are multiple landmarks including the corpus callosum near the cortex, the internal capsule running parallel along the trajectory, and the anterior commissure, which runs medial–lateral near the target. Like the procedure described for targeting M1, all cortical and subcortical targets should be confirmed with electrophysiological recordings. For the nucleus accumbens, the characteristic changes in the recordings are marked by large amplitude neuronal firing when passing through the corpus callosum, quieting of neuronal firing in the internal capsule, bursting in the caudate and putamen, and large amplitude firing again if electrodes enter into the anterior commissure. In general, most neuronal structures have a characteristic firing pattern when the electrophysiological recordings are monitored with an audio speaker. In addition, the borders between neuronal structures also provide a differential in signal characteristics and should be used to confirm correct insertion trajectories (Rezai et al., 2006). For any cortical or subcortical structures, as much information about the relative anatomical layout, significant landmarks, and characteristic electrophysiological properties should be known before surgery.

4.2.1.3 Targeting in humans   The need for accurate placement of electrodes in deep brain stimulation (DBS) procedures has sparked the development of many clinical targeting systems that have advanced capabilities (Perlmutter and Mink, 2006). The fundamental functioning of these systems is the same as those described in the M1 and NAcc examples above. However, the technology has developed over the years to improve safety, accuracy, and fusion of information to ensure the best clinical outcome for the patient. Clinical systems such as Stealth from Medtronic allow for the importation and manipulation of multiple imaging modalities that include MRI and CT scans, which are routinely performed in advance of the surgery. It is important to acquire these images within hours of the surgery because changes in intracranial pressure, blood pressure, and movement of the brain in the CSF can cause variance between the static image and the desired real-time targets needed for DBS surgery. Millimeter or submillimeter accuracy is needed for electrode placement in these human applications and small variances in the targeting plan and imaging can cause complications in outcome efficacy. To overcome these issues, companies like Medtronic have developed software (Figure 4.4a) that allows for precise image fusion of multiple data sets so that an aggregate of information can be obtained. Because alignment of brain structures is not guaranteed among the imaging modalities (due to resolution, etc.), advanced image correlation algorithms enable precise manual or automatic image fusion. Lastly, surgeons are aided in the incorporation of interactive brain atlases and a Talairach grid to match imaged structures with known anatomical landmarks. Altogether, surgeons are able to identify critical landmarks, derive target coordinates, and map trajectories.

In addition to commercial systems, custom systems such as the FGATIR exploit progress in imaging. The FGATIR sequence of imaging offers significant advantages over both standard T1-w and T2-w FLAIR imaging for targeting utilized for DBS in movement disorders (Sudhyadhom et al., 2009). In Figure 4.4b, an example FGATIR image is shown overlapped with a stereotaxic atlas. One unique feature of the FGATIR is that it can also be used to mark electrophysiological indicators as a function of depth while driving the electrode into the target. Such markers, as indicated by the green dots, can be used to help define areas of normal or abnormal neuromodulation as well as borders between structures. In addition, the correspondence between the observed electrophysiology, electrode position, and subcortical atlas boundaries can be evaluated to gain confidence in the targeting or to make adjustments.

4.3 Surgical methods for implantation

Successful implantation and recovery from the surgical procedure of inserting microelectrode arrays is predicated by extensive preparation of materials, instruments, and the operating environment. These preparations must be coupled with extensive mental and physical practice of the surgical methods. A priori development of strategies to mitigate risks and sets of alternative courses of action are essential for the smooth operation of the procedure. Before any surgical procedure, it is advantageous to perform a "dry run" of all of the steps to anticipate any potential problems and to work out any complications that could take valuable time in the operating theater. Recovery from surgery is almost always improved with minimal duration and minimally invasive procedures. Steps should be taken to reduce, refine, and streamline surgical procedures. The surgical methods presented here are a guideline for chronic implantation of multiple designs of

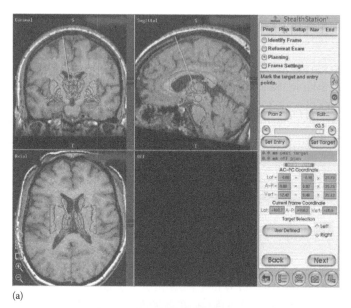

(a)

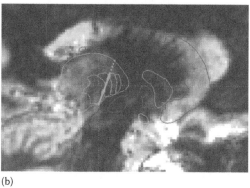

(b)

Figure 4.4 Targeting software for human deep brain stimulation. (a) Medtronic stealth system showing fusion of imaging, a deep brain atlas (gray outlined structures), and a target trajectory plan (gray line). (Reprinted with permission from Halpern, C. H., U. Samadani, B. Litt, J. L. Jaggi, and G. H. Baltuch, Deep brain stimulation for epilepsy, *Neurotherapeutics*, 5 (1), 59–67, 2008, doi: http://dx.doi.org/10.1016/j.nurt.2007.10.065.) (b) Fgatir system that allows for correlating electrophysiological measurements (gray dots) with imaging and atlas (shades of gray outline) information (Sudhyadhom, A., I. U. Haq, K. D. Foote, M. S. Okun, and F. J. Bova, A high resolution and high contrast MRI for differentiation of subcortical structures for DBS targeting: The Fast Gray Matter Acquisition T1 Inversion Recovery [FGATIR], *Neuroimage*, 47, Suppl 2 (19362595), T44–T52, 2009. doi: papers://7964EF6A-BE0A-4F2B-9048-9B5C50D5342A/Paper/p10057.)

58

electrode arrays. They should be adapted and improved over time with advances in surgical technique and technology.

4.3.1 Surgery preparation and sterilization

Surgery preparation begins with the collection and sterilization of all of the necessary tools and instruments needed to conduct the procedure. At least 1 to 3 days before the procedure, all electrodes should be prepared and checked for their quality of fabrication, integrity of the ground wires, and impedance to ensure that they are in the normal ranges. Backup electrodes should be included in case one or more break, become unsterile, or are not inserted properly. In addition to electrodes, all surgical instruments should be wrapped, labeled, and sterilized before surgery. Materials needed for surgery should also be collected in advance. These include monitoring equipment, adhesives, batteries, and fluids. The day before the surgery, all major equipment (stereotaxic instrument, microscope, operating table, etc.) should be wiped down with alcohol or other cleaning agent to minimize the risk of infection. All instruments in contact with the surgical site should be sterilized with either gas or temperature/pressure techniques. Depending on the type of instrument or material, gas sterilization may be preferable over autoclave methods. Figure 4.5 shows examples of table-top versions of these instruments that can be acquired for small or large neuroprosthetics labs. Typically, instruments with fine tips, fragile electrodes, or materials affected by water vapor would undergo gas sterilization. All other materials and equipment would undergo sterilization in an autoclave. Most academic institutions and medical facilities have core sterilization equipment that is available for use. If access to these facilities is difficult, timely, or expensive, there are a variety of tabletop sterilization equipment that can be purchased. These pieces of equipment are appropriate for laboratories that do not have a high-throughput of subjects for implantation.

The following materials provides a partial list of the major equipment, instruments, and materials used for the implantation of microelectrode arrays. This list should be augmented or modified with each particular surgeon's preferences or should be tailored for a particular procedure.

- Gas sterilization
  - Cauterizer (batteries removed)
  - Burs: assortment (round head/105, square head/110, 0.5 mm head, have some spares)
  - Drill bit (2)
  - Drill tap: handheld and pin vise size
  - Trephine (2)
  - Screwdriver for screws (4)
  - Pin vise for drill bit and tap (2)
  - Super glue

(a)

(b)

Figure 4.5 Tabletop sterilization equipment. (a) Ethylene oxide (EtO). (b) Autoclave.

- Petri dishes (1), for acrylic mixing
- Stereotaxic pointer array and dummy array
- TI screws (12)
- Long screws for grounding (2)
- Allen wrenches for stereotaxic equipment
- Stereotaxic equipment: stereotaxic arms (2)
- Stereotaxic equipment: array holder for array holding arm (2) and long holder
- Headstages and amplifier connector cable (2, need for both arrays)
- Drill
- Cups (5) to hold: saline, alcohol, peroxide, water, and leftover liquids
- Alligator wire (with extra clasps) for grounding
- Surface mapping electrodes (pair plus single spare)
- Electronic microdrive
- Autoclave
  - Scalpel handle
  - Dura cutters, straight, small
  - Dura cutters, curved
  - Dura hook
  - Cotton rolls
  - Cotton swabs
  - Cotton gauze (at least four bags)
  - Needle holder
  - Towel clamps (8)
  - Scissors: small and medium
  - Hemostats, small (2)
  - Curved tip, small hemostats (5)
  - Instrument towels, 24 × 24 in. (4)
  - Fine serrated tweezers
  - Fine forceps, smooth and pointed
  - Fine forceps, angled, smooth, and pointed
  - Angled, serrated forceps
  - Gently curved tip, smooth forceps
  - Medium serrated forceps
  - Bone rongeur, fine
  - Bone clipper, fine
  - Spatula for scraping
  - Small dental elevator
  - Spatula for cement
  - Medium, curved hemostat
  - Surgical microscope handle covers
  - Cement mixing cup
  - Blunt tip 18 Ga needles (at least 12)

- Presterilized or not sterilized
  - Cautery batteries
  - Computer (with mouse pad, speakers, audio cable, and BNC audio adapters)
  - Amplifier batteries (make sure it is fully charged)
  - Saline, isopropanol (alcohol) 99%, hydrogen peroxide, iodine
  - Cold saline for mapping
  - GentaCement (2)
  - Bone wax
  - Gel foam
  - Eyelube
  - Antibiotic ointment
  - Disposable drapes: enough for sterile field
  - Sterile gloves
  - Sterile markers
  - Suture: for closing
  - Scalpel blades
  - Syringes, 1 cc (at least 12), 5 cc, 60 cc
  - Needles: 18, 25, 27, 30 gauge
  - Hair clippers
  - Goggles (2)
  - Stereotaxic equipment: stereotactic frame/base, with rail
  - Microdrive with insertion arm
  - Drill footplate/power supply
  - Extension cords/power strips (3)
  - Calculator
  - Camera
  - Surgical documents
  - Atlas, skull, pictures of insertion paths, mapping stimulation parameters, array layout
  - Chemical heating pads
  - EM meter
  - Array connector caps (5)
  - Dissecting microscope
  - Hot bead sterilizer
  - Scissors and tape
  - Stimulator
  - Electrolyte

4.3.2 Preparation of the operating theater

The morning of surgery, the operating theater should be set up for the procedure. Special attention should be made to the layout of electrophysiological equipment, instruments, operating table, and vital monitoring equipment. The items should also be arranged according to their sterility. Typically, all sterile items are on or near the surgical worksite and all nonsterile items are located farther away. An example of this layout is shown in Figure 4.6a. Here, the surgeon on the right has direct access to a sterile field where all critical instruments and supplies for the surgery are placed. Outside this sterile field, an assistant (shown on the left) can

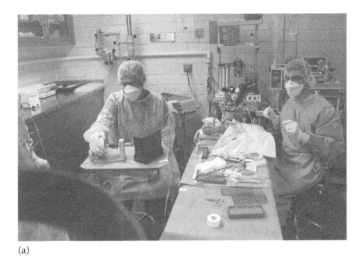

(a)

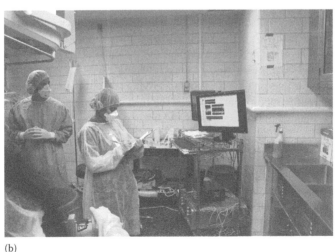

(b)

Figure 4.6　Arrangement of the operating room theater. (a) Layout of the sterile surgical field, sterile instruments/supplies in close proximity, and nonsterile equipment used by assistants. (b) Monitoring equipment used for electrophysiological verification of targets.

63

operate nonsterile equipment such as a microdrive for advancing electrode arrays into brain tissue. In addition to surgical equipment and supplies, it is typical to also have electrophysiological monitoring equipment (Figure 4.6b) in the operating theater, which can be observed by the surgeon and assistants. Such equipment is used to validate the targets of implantation by assessing the signatures of neuronal firing in brain substructures. Such approaches are commonly used in deep brain stimulation (DBS) surgical procedures (Rezai et al., 2006).

4.3.3 Surgery preparation: Physiological monitoring and anesthesia

Preparation for surgery involves readying the subject for anesthesia and physiological monitoring. Although the exact details of surgery preparation will differ between animals and humans, the general set of steps is similar for most CNS neuroprosthetic implant procedures. These steps should serve as a guide; however, the investigator or surgeon should tailor the procedure for each specific subject and application. The investigator needs to first determine if conscious sedation or general anesthesia will be used for the procedure (Venkatraghavan et al., 2010). The choice depends on the type of electrophysiology or brain mapping the investigator seeks to perform during the procedure. The effects of anesthesia are not uniform across the brain or neuronal pathways. Certain methods of anesthesia may block pathways that may be necessary for recording or stimulation. A classic example of this is choosing inappropriate anesthesia when one is trying to perform cortical stimulation to evoke muscular responses in the limbs. Certain anesthetics may block cortical–spinal and neuromuscular pathways. In addition to the pathways, anesthetics may affect the excitability of neuronal networks and the activity monitored during the implant procedure may not be representative of known firing patters. Such information could bias targeting or interpretations of the underlying state of the neuronal network in question. Once the appropriate anesthetic regimen has been selected, the next step is to assemble the monitoring equipment and begin drug administration.

Because CNS implants are often multi-hour procedures, surgery usually begins in the morning with the administration of anesthetic. Common anesthetics for conscious sedation include propofol and opioids such as fentanyl. For general anesthesia, isoflurane, and constant rate infusions of ketamine can be used. Next, the subject is first prepared for intubation (in the case of general anesthesia). The subject can be intubated with an endotracheal tube to establish an airway using standard laryngoscopy technique. In animal subjects, where intubation may be difficult, atropine can be administered to reduce salivation. Under anesthesia, the subject should be monitored using an electrocardiogram to assess heart rate, arterial blood pressure, oxygen saturation, and end-tidal CO_2. The complex interplay of drugs and physiological parameters is extremely important for neuroprosthetic implant procedures. Stable planes of anesthesia and physiological parameters are desirable because they can affect the ability to target brain structures and perform intraoperative electrophysiological measurements. One of the most important of these physiological parameters is end-tidal CO_2, which should be maintained at approximately 3.8%. Improper management of ventilation will result in an excess of CO_2, triggering an autonomic vasodilation response in the subject to rid the body of this excess. Vasodilation also leads to increased intracranial pressure, which can distort the positioning and shape of the brain within the cranium. Such distortions will cause inaccuracies in targeting. In addition to ventilation, body temperature should be closely monitored and maintained with either a water circulating or chemical heating pad. These types of heating pads are desirable because they are less likely to introduce electrical noise into electrophysiological monitoring during the procedure. Improper maintenance of body temperature will lead to either too deep or too shallow anesthesia and complicated targeting and physiological measurements. In addition to physiological monitoring, intravenous lines should be placed for the delivery of antibiotics and fluids during surgery. Both will assist with recovery especially during long procedures (those in excess of 2–4 h). The time and duration of surgery should be continuously monitored and effort should be made to reduce surgical time.

4.3.4 Stereotaxic frame

The stereotaxic frame is usually placed on the subject's skull under local or general anesthesia. Many of the procedures for modern stereotaxic frames were developed in 1908 by two British scientists; Sir Victor Horsley, a physician and neurosurgeon, and Robert H. Clarke, a physiologist (Picard et al., 1983). Many variations of the original device began to appear around 1930 in animal studies and modifications to human neurosurgery. The general approach for placing a subject into the frame has remained unchanged. It consists of anchoring the frame to the skull using a variety of landmarks. For human deep brain stimulation stereotaxic neurosurgery, as shown in Figure 4.7a, the frame is affixed to the skull using four screws and a local injectable anesthetic. For animals, the procedure is more complicated and commonly consists of using a Kopf stereotaxic frame (http://www.kopfinstruments.com) in which there is a bite bar, ear bars, and infraorbital stabilizers. The subject is placed on the frame by centering the head and incisors on the bite bar. Next, the ear bars are inserted into the ear canals. Special attention needs to be made such that the points on the ear bars are directly centered in the middle of the canals and not pressing against other soft tissue. This can be confirmed with the ears laid flat and symmetrically around the ear bars. Improperly placed ear bars can cause increased fluid buildup, swelling, or intracranial pressure due to the ear bars pressing against bone or skin. Once the subject is centered and the adjustable components of the stereotaxic device are locked down, the subject should be checked for physiological parameters such as respiratory rate, discomfort, and overall welfare. The last step of applying the stereotaxic frame and preparation of surgery consists of sterilizing

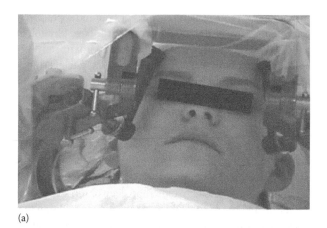

(a)

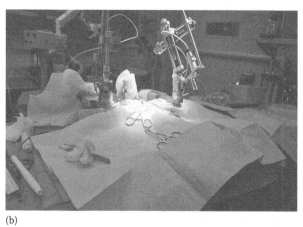

(b)

Figure 4.7 Stereotaxic frames; (a) human deep brain stimulation (DBS) frame and (b) Kopf stereotaxic frame for primates.

the head and frame. This can be achieved using a combination of iodine solution, chlorhexidine, and alcohol. Most stereotaxic neurosurgeons use multiple swipes of an iodine solution to thoroughly saturate the surgical site. Iodine solutions are most effective when allowed to soak on the site for at least 3 minutes. During the sterilization procedure, the solution is worked from proximal to the surgical site to distal, moving debris away from where the incision with take place. Once the iodine has been allowed to soak, excess can be cleaned off using alcohol. The surgical site can then be draped to cover all areas not involved in the procedure. The final surgical setup and draping should allow the surgeon to clearly observe all extremities (hands, arms, legs, and feet). Access to these limbs is particularly important when conducting brain mapping procedures where movement is being evoked. Likewise, for sensory or motor-evoked studies, movement of the limbs can be used to produce neuronal responses. All the aforementioned methods can be used to add confidence to the stereotaxic neurosurgical technique and targeting.

4.3.5 Surgery and defining the layout

Surgery commences after the surgeon(s) have scrubbed into the procedure to ensure sterile techniques. For brain implants, the first step is to perform an incision in the vicinity of the targeted electrode insertion site. It is always beneficial for the subject to create the most minimally sized incision necessary to complete the procedure. This reduces the wound margin and the chance of infection. After the incision, the skin should be reflected back and secured with hemostats to expose the skull, as shown in Figure 4.8. A clean surgical site reduces complications; therefore, supporting tissues such as the periosteum on the skull are removed. Hydrogen peroxide can then be applied to the skull to more clearly expose the sutures needed for targeting and for cleaning excess blood and tissue. Once the skull has been cleaned, it can be checked for any bleeding. Using a small cautery, sites of localized bleeding can be stopped by quick taps of the instrument. Maintaining a clean and dry skull in the vicinity of the implant is one of the aspects of maintaining the longevity of neural implants during chronic procedures. Recording sites that are leaking blood or cerebral spinal fluid can be breeding grounds for bacteria especially because they are in enclosed, dark spaces and at physiological temperatures. At every step of the surgical procedure, effort should be taken to minimize the effects of these aspects. With the skull clean and dry, it is appropriate to use a sterile pen to mark the general layout of

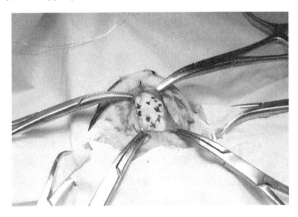

Figure 4.8 Initial incision and reflection back of the scalp. Here, the periosteum has been scraped from the bone and the skull has been cleaned with hydrogen peroxide. The top of the skull has also been marked with a sterile pen to indicate the surgical layout of hardware.

the craniotomy and all related hardware. Special care should be taken at this point to make sure that all of the hardware will fit in the surgical site. This includes all screws, electrodes, ground wires, or percutaneous connectors. It is a worst-case scenario to have to rework the implant layout during the surgical procedure because it adds surgical time and often leads to suboptimal, rushed decisions. At the time of marking the surgical layout, the landmarks of the bregma, lambda, or other sutures can be identified if they will be used in the stereotaxic targeting.

Many of the marks on the skull in Figure 4.8 are for indicating the position of anchoring and grounding screws. These screws will hold the entire cranial implant assembly in place and are affixed to the skull. The layout of the screws should be such that they form a structure that resists forces in all directions. Alignment of screws in a square pattern can achieve this feature. Once the basic set of anchoring screws has been marked, additional screws for supporting other hardware or the electrode during insertion can be added. A good rule of thumb is to use at least four to six screws for anchoring and support. Of this set, one screw will need to be used for a ground location. As an alternative, an additional screw can also be added to serve as the ground. Once the screw positions have been marked, the next step is to begin drilling. It is advantageous to place anchoring screws before opening the cranium to prevent any damage to the brain (Figure 4.9). The process of putting screws into the skull consists of drilling a pilot hole, tapping the hole, and then inserting the screw. The choice of pilot drill bit size, tap size, and screw size should be tailored to the subject. Beneficial outcomes can be achieved with titanium screws that are not self-tapping and are well matched in length to the thickness of the subject's skull. Self-tapping screws should be avoided because the sharp point on their tip could abrade the brain or dura surface. During the drilling and tapping of the screw holes, the surgeon should pay attention to the anatomy of the skull, which consists of a hard top and bottom table, and a middle layer with lesser density. Using

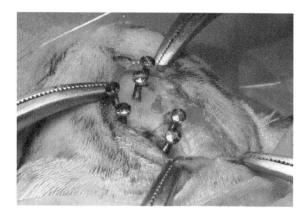

Figure 4.9 Initial set of anchoring screws.

68

a hand drill, the surgeon should be able to feel the differences between the layers and they can help to guide how much pressure to apply during the procedure. Puncturing the skull, dura, and brain should be avoided at all times. The thickness and layers of the skull also guide the number of turns to complete while driving the screws during insertion. Skull thicknesses can range from 1.5 to 5.0 mm. The surgeon should be able to feel the resistance on the screw change while passing through the top, middle, and bottom layers of the skull. The tip of the screw should not extend or extend very minimally past the bottom table of the skull. For grounding screws, the tip must be in contact with the CSF or proper grounding will not be achieved.

Once the screws are placed, the next step is to begin drilling the craniotomy to expose the dura and brain cortex. Because the procedure for opening the cranium can produce vibrations and evoke autonomic responses, it is common to slightly deepen the anesthetic plane before drilling. At this point, a decision needs to be made as to whether a burr hole or window will be opened in the skull. Typically, burr holes are much smaller than windows and can be used to accommodate smaller electrodes. They are the standard of care for DBS-style surgeries because they minimize the invasiveness. In contrast, window procedures open a large portion of the skull to expose not only the desired target but also related anatomy that can be used to assist targeting. The window approach is also often used to place electrocorticogram (ECoG) grids, which can be difficult to deploy in smaller burr holes; however, new methods are being developed to overcome this problem (Thongpang et al., 2011). The process of drilling a burr hole or window is quite different. The burr hole can be achieved with a single large drill bit to open the cranium, as shown in Figure 4.10. For the

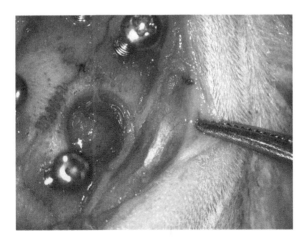

Figure 4.10 Craniotomy used to implant chronic microelectrode arrays. The dura is cut in this example and reflected back over the bone (top of craniotomy). The exposed brain and vasculature is in the center.

69

window procedure, an outline is typically made to mark the window site. Next, four holes are drilled in the corners of the window. Last, a bone saw is used to cut out the outline of the window. The entire bone flap can then be removed to expose the dura and brain.

Once the cranium is opened, the dura will be exposed. Extradural electrodes can be applied directly to the surface, but for subdural electrodes, the dura will need to be opened. This is commonly achieved with a dura hook or a fine needle (25 g) with a bent tip, which can be used to grasp and pull up the dura such that a scalpel or microscissor can be used to cut the tissue. Incisions are made in the dura so that it can be opened and reflected back (as shown in the top portion of the craniotomy in Figure 4.10) to expose the surface of the brain.

4.3.6 Cortical and subcortical mapping

Once the dura has been opened, cortical mapping can commence to confirm stereotaxic targets. Brain mapping is often used to define motor or sensory areas of the target neural structures, and it involves delivering current via a bipolar electrode (Figure 4.11a) to evoke movement in the extremities or perception of sensation. Before beginning mapping, the desired electrode implant positions should be defined with rough estimates of where the borders of the electrode array will ultimately end up. The mapping procedure can begin at these extremity points to define the boundaries. Stimulation can be delivered via a commercial, programmable stimulator such as the A-M Systems Isolated Pulse Stimulator Model 2100 as shown in Figure 4.11b. Typical stimulation parameters include frequencies around 50 Hz (20 ms period), amplitudes ranging from 1 to 7 mA, and pulse widths in the range of 100 μs. Stimulation parameters will depend on the type of bipolar electrode, size of the electrodes, and the target. For each stimulation parameter set, the subject should be visually inspected for movement of the extremities or signs of sensory perception. Visual signs of stimulation activation should be coupled with knowledge of the motor or sensory homunculus to confirm the mapping outcomes. For example, confirming a serial order of hind limb, trunk, forelimb, and face activation in M1 would provide good confirmation that the motor cortex was localized and mapped.

4.3.7 Electrode driving

When high confidence in the neural target of choice has been achieved, it is time to begin the electrode implant procedure. There are four main types of microelectrode arrays that are used in practice and they include microwires, floating microprobe arrays, Blackrock (Utah arrays), ECoG arrays, and silicon arrays (Michigan electrodes). The procedures for implanting each type are different and they are outlined below.

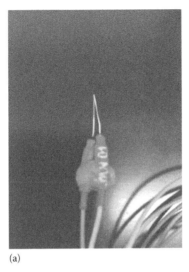

(a)

(b)

Figure 4.11 (a) Example of a bipolar electrode used for cortical mapping. (b) A-M Systems programmable stimulator.

4.3.7.1 Microwires Implantation of microwire arrays involves extensive use of a micropositioning system such as those that can be obtained from Kopf Instruments (Figure 4.12a). A micropositioner is useful to achieve very slow and precise insertion, which is believed to exert the least amount of trauma to the tissue and vasculature (Nicolelis, 1999; Nicolelis et al., 2003). In addition to minimal tissue damage, the use of a micropositioner will allow simultaneous recording during the driving of the electrode to confirm neuronal targets and borders between structures. Classic landmarks that are often used during driving of electrodes include layer V of the cortex, the corpus collosum, and fiber tracts such as the internal capsule and anterior commissure. The process of implanting microelectrode arrays begins by attaching the array to the recording system and then attaching the assembly to the stereotaxic drive arm as shown in Figure 4.12b. The array should then be positioned in the middle of the craniotomy and lowered slowly until the tips of the electrodes just touch the surface of the brain. The electrode array can be grounded via an external wire connected to the ground screw. Special care should be taken

71

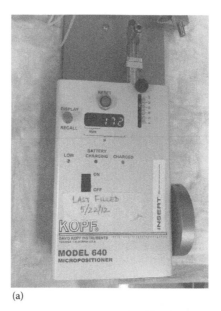

(a)

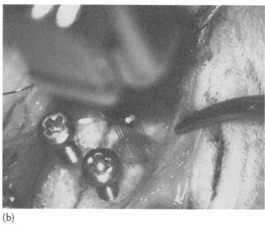

(b)

Figure 4.12 (a) Kopf micropositioner system used for implanting microelectrode arrays. (b) Implantation of a 16-channel microelectrode array. The electrode is affixed in a recording headstage that is attached to a stereotaxic arm.

to ensure that all the electrodes are straight and none are caught on tissue around the edges. Electrodes that are "hung up" on tissue will cause serious complications including hemorrhage and compression of the brain if they are not freed before the driving process. The method for checking these aspects should be performed under a microscope. Once it is determined that the electrode is free and clear to drive, the assembly should be advanced 100 µm. The surgeon should check that there is no compression and dimpling on the surface of the brain. The electrode array should penetrate smoothly into the tissue. Sharpened tip electrodes can also facilitate this process but they are not necessary. The array can then be slowly advanced into the tissue in increments of 100 µm. Observations of signals across the array should be logged at each depth and time. These observations can be used to infer the 3-D spatial geometry of neural structures related to the target of choice. Once the target has been achieved, the electrode array can be anchored with a small amount of bone cement to a nearby screw.

4.3.7.2 Microprobes (FMA electrodes) The insertion of microprobe arrays involves the use of a vacuum pump (Figure 4.13a) and is similar to the insertion into microwire arrays but is a much more delicate procedure (Musallam et al., 2007; Mollazadeh et al., 2011). First, the FMA should be gently removed from its shipping box using fine tweezers. With the vacuum pump metal shaft fixed into the micropositioner apparatus, turn on the vacuum pump. The FMA should be gently held using the fine tweezers and centered on the vacuum shaft. The electrode will be helped by the vacuum. Do not turn off the vacuum pump at any time or the array will fall off and be damaged. Note that recording during implantation of FMA arrays is not possible due to the noise from the vacuum pump. Next, the FMA should be positioned in the center of the craniotomy. Make sure the wire bundle is close to one of the screws. That screw will be used to fix the wire bundle so that the array does not move once the vacuum is turned off. Start lowering the electrode into the cortex by turning the micropositioner dial. To hold the wire bundle in place, put a small drop of cyanoacrylate gel onto the wires and screw so that the wire is glued to the screw. Once the wire bundle is glued, turn off the vacuum pump and release the pressure valve. After the vacuum is released, the vacuum shaft can be removed. To seal the implant site, cut small pieces of gelfoam and soak it in sterile saline. Cover the craniotomy with gelfoam pieces on and around the FMA so that the whole space is filled with gelfoam. Finally, mix a very thick layer of cranioplast and put it on top of the craniotomy. Make sure the cranioplast does not flow into the craniotomy. Place additional layers of cranioplast to anchor the FMA properly, making sure no cranioplast gets on the grounding screw.

4.3.7.3 Blackrock (Utah electrodes) The procedure for implantation of Blackrock arrays contains very few steps due to the methodology for injecting the array into the tissue. The surgery relies on the use of a pneumatic injector, as shown in Figure 4.14a. Using this tool, the array is rapidly injected into the cortical tissue to overcome forces between the tissue and the array so that injury is minimized (Nordhausen et al., 1994; Maynard et al., 1997; Rousche and Normann, 1998; House et al., 2006). Unlike the microwire and FMA procedures, the first step for the Blackrock array insertion is to position, by hand, the array directly on

73

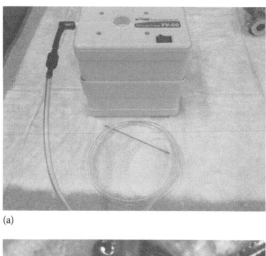

(a)

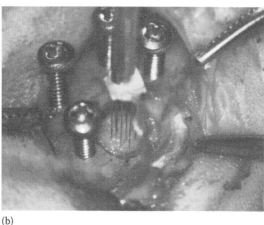

(b)

Figure 4.13 (a) Vacuum inserter used for implanting FMAs. (b) FMA attached to vacuum inserter while being implanted into the brain.

the surface of the cortex. Special care must be taken to prebend the wire bundle such that the array sits flat on the cortex. Once the array is positioned, the next step is to attach the pneumatic injector to the stereotaxic arm. The injector piston is then aligned directly over the top of the array. The tip of the pneumatic injector must hit the top of the Blackrock array squarely to distribute the force equally across the entire array. If this does not occur, only part of the array will be implanted; leaving multiple shanks out of the tissue. To control the amount of force and excursion of the injector, multiple settings can be used. Typical settings include a pressure of 20 psi, pulse width of 6.0, and a 1.0-mm insertion piston. Once the array is tapped into place, the craniotomy is sealed in a similar manner

(a)

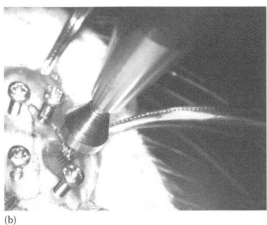

(b)

Figure 4.14 (a) Pneumatic injector used for implanting Blackrock arrays.
(b) Aligning of pneumatic injector before implantation of a Blackrock array.

as the FMA electrodes. Cut small pieces of gelfoam and soak it in sterile saline. Cover the craniotomy with gelfoam pieces on and around the array so that the whole space is filled with gelfoam. Either cranioplastic cement or replacement of a bone flap can be used to seal the insertion site.

4.3.7.4 ECoG Unlike the implantation of microelectrode arrays, the procedure for implanting an ECoG array typically involves large craniotomies. These are needed because the spatial arrangement of most clinical ECoG grids has 1 cm center-to-center spacing of 0.4-mm electrodes, as shown in Figure 4.15. Once the craniotomy and dura are opened, the insertion of the electrodes consists of laying the grid directly on top of the cortex. Special care should be

75

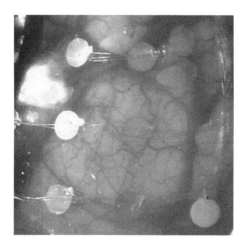

Figure 4.15   Placement of a clinical ECoG grid on the cortex.

taken to ensure that all electrodes are in contact with the brain surface. Failure to do so will result in poor quality recordings. In addition, it is desirable to have all of the electrodes in contact with neural tissue. In Figure 4.15, the top center and left electrodes are directly located over the vasculature. These sites will likely not provide high-quality neural recordings. Once the electrodes are placed, their wire bundles are sutured to either the dura or anchored to the skull to prevent movement. A bone flap or other device such as a Stimloc lead kit (Medtronic, Minneapolis) can be used to close the craniotomy depending on the size and method of implantation.

4.3.8 Grounding, closing, and postoperative recovery

The last step in the surgical implantation of electrode arrays is to ensure proper grounding, closing, and postoperative recovery. Depending on the electrophysiological setup, the grounding procedure may differ. In the simplest case, both the ground and reference electrodes can be attached to the same screw located in the skull. This is called "single ended" recording. If multiple electrode arrays are used, they must all be tied to the same ground point to ensure the best recordings. Do not tie a separate ground to each electrode array as this will allow more noise and artifact to enter the system. In situations where stimulation is being used or if it is desirable to make the system more robust to noise, a "differential" setup can be used where the ground wire is connected to a screw in the skull and a reference wire is implanted in the brain near the recording sites of the array. In all configurations, it is essential that good electrical contact is made between the ground/reference wires and their target sites. Improper connectivity will introduce noise and artifact.

After all the connections are checked, it is time to close the procedure. It is essential that the craniotomy is completely sealed so that excess blood or CSF does not leak. Leakage of these fluids will lead to increased risk of infections, softening of the skull, and total loss of the implant. Craniotomies can be sealed with medical grade cyanoacrylate, gel foam, silicone polymer, and cranioplastic cement. With the skull clean and dry, the scalp should be closed around the implant with sutures or staples. Antibiotic ointment should be applied around the wound margin to prevent infection.

Postoperative recovery consists of moving the subject to a site where close monitoring can take place. This includes looking for signs of alertness, ambulation, eating, drinking, and proper body temperature. If the subject is in an extended period of a drowsy state, refuses to ambulate/eat/drink, or has low body temperature, there are likely secondary recovery issues that must be addressed immediately. A regimen of postsurgery medications should be given for pain and antibiotics can be administered to prevent infection. The first 24 h postsurgery are the most critical for good recovery and the surgical and support team should be attentive and proactive to any issues that may arise.

4.4 Surgical methods for perfusion

Once subjects have completed their participation in the study, it may be desirable to analyze the tissue around the electrode array implants to determine targeting, signal quality, and tissue responses. To extract brain tissue, the subject must first be perfused. Before perfusion, subjects were deeply anesthetized and transcardially perfused with a fixative solution containing 10% formalin in 0.1 M phosphate buffered saline (PBS) (pH 7.4). Using the procedure shown in Figure 4.16, an abdominal incision is made

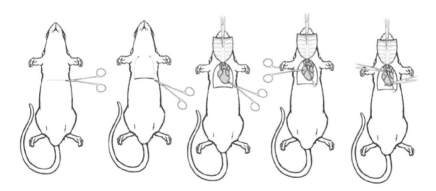

Figure 4.16 Surgical method for transcardial perfusion. (Adapted from Gage, G. J., D. R. Kipke, and W. Shain, Whole animal perfusion fixation for rodents. *J Vis Exp* (65), 2012. doi: papers2://publication/doi/10.3791/3564.)

77

along the sternum and the chest cavity is opened to expose the heart. This is achieved by first feeling for the bottom of the sternum. Using tweezers, the skin can be pulled up and cut to create a small hole at the bottom of the sternum. Next, cut left and right along the ribcage toward the arms and peel back the sternum rostrally. This step will expose the diaphragm, which needs to be cut through to expose the heart and lungs. A needle is inserted into the left ventricle and a hemostat is used to clamp the heart with the needle inserted into it. A perfusion pump is then used to pump 0.9% PBS at a very slow flow rate into the subject's heart. At the start of the pumping, a slit is made in the right atrium to flush out the perfusion fluids. Special care should be taken to ensure that the needle is located in the aorta and pumping fluids throughout the body. If the heart septum is punctured or if the needle is in the pulmonary artery, the lungs will begin to fill with fluid and the needle should be repositioned. Once the fluid from the atrium runs clear and the liver shows signs of blood clearance, the pump is switched to deliver 10% formalin. When the formalin is delivered, cross-linking of the tissues will begin immediately. Obtaining twitching of these tissues especially in the extremities are a sign of good perfusion.

4.5 Surgical methods for explantation

Upon perfusion, the head of the animal harboring the electrodes should be detached from the rest of the body using a guillotine. Place the blade just behind the ears and make one swift cut. As shown in Figure 4.17, insert scissors in the foramen ovale and make cuts to the left and right being careful not to cut into the brain. Next, break off the top of the skull and use a spatula to scoop out the brain by severing the cranial nerves on the bottom and the optic nerves in the front. Lastly, place the brain in a formalin-filled cup for 24 h.

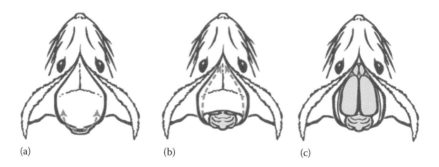

(a) (b) (c)

Figure 4.17   Extraction of the brain from the skull. (a) Insertion point for scissors, (b) continue cutting along dashed lines and break off top of skull, and (c) scoop out the exposed brain with a spatula. (Adapted from Gage, G. J., D. R. Kipke, and W. Shain, Whole animal perfusion fixation for rodents. *J Vis Exp* (65), 2012. doi: papers2://publication/doi/10.3791/3564.)

Exercise

1. Perform a presurgical targeting analysis for an animal subject. Indicate the stereotaxic coordinates of the target, the insertion sites, and location of extracranial hardware for implantation. Use a stereotaxic atlas to show the trajectory of the implant. Draw the extracranial location of all hardware.

Chapter 5 Quantifying long-term electrode performance

Should you find yourself in a chronically leaking boat, energy devoted to changing vessels is likely to be more productive than energy devoted to patching leaks.

Warren Buffett

Learning objectives

- Understand the mechanisms of why neural–electrode interfaces perform well and why they fail.
- Understand the variability over time of electrode performance in neuro-prosthetic applications.
- Decide what strategies can be used to promote long-term electrode performance.

5.1 Introduction

The development of clinically viable microelectrode arrays for humans has produced multiple engineering designs and neurophysiological requirements that are believed to be necessary for facilitating high performance (Jackson and Zimmermann, 2012). These include the ability to target and access the activity in ensembles of neurons located in cortical and deep brain structures for treating a variety of neurological problems, including paralysis (Ethier et al., 2012; Hochberg et al., 2012; Collinger et al., 2013), stroke (Iosa et al., 2012), movement disorders (Andrews, 2010; Rouse et al., 2011), epilepsy (Morrell, 2011; Truccolo et al., 2011), and neuropsychiatric disorders (Maling et al., 2012). Not only is the short-term access of these neurons important, but the microelectrode arrays should be able to sense and stimulate targeted neurons for the life of the implant and patient, which could be on the order of 70+ years. From a materials science perspective, there are differences in the design considerations between sensing and stimulation, but in general, it is desirable to have stable electrodes that do not corrode (Sanchez et al., 2006; Prasad et al., 2012), produce stable impedance (Ward et al., 2009; Prasad and Sanchez, 2012), and are robust after the repeated delivery of current through them (Johnson et al., 2005; Otto et al., 2005, 2006; Koivuniemi and Otto, 2011; Lempka et al., 2011). A variety of materials, including tungsten, platinum, platinum-iridium, and iridium oxide coatings, have been used to improve the reliability of recordings (Geddes and Roeder, 2003; Cogan, 2008; Ward et al., 2009). Closely coupled to the abiotic material aspects are the biotic responses to chronically implanted microelectrode arrays. The "biotic" effects include the disruption of the blood–brain barrier (BBB) (Prasad et al., 2011; Karumbaiah et al., 2013; Saxena et al., 2013), tissue inflammatory response involving astroglial and microglial reactions occurring at or around the implanted site, and macrophage recruitment to the implanted site (Schmidt et al., 1976; Stensaas and Stensaas, 1978; Edell et al., 1992; Kam et al., 1999; Turner et al., 1999; Szarowski et al., 2003; Biran et al., 2005, 2007; Lee et al., 2005; Polikov et al., 2005; McConnell et al., 2009; Winslow and Tresco, 2010; Thelin et al., 2011; Prasad et al., 2012).

The "abiotic" effects include the changes occurring at the electrode recording sites such as corrosion and insulation delamination and cracking that alters the electrochemical properties of the electrode recording surface area (Geddes and Roeder, 2003; Patrick et al., 2011; Prasad and Sanchez, 2012; Prasad et al., 2012; Streit et al., 2012; Kane et al., 2013). Both these effects are dynamic in nature, occur concurrently, and cannot be isolated from one another. Therefore, high-performance arrays should produce minimal tissue reactivity, including disruption of the blood–brain barrier (Prasad et al., 2012), glial response (Frampton et al., 2010; Winslow and Tresco, 2010; Prasad et al., 2012), and neuronal damage (McConnell et al., 2009). Finding the optimal balance of all of these design considerations is challenging, and multiple array solutions have been proposed, including microwire arrays, floating planar silicon arrays, floating 2-D silicon arrays, and floating microwire arrays (Drake et al., 1988; Rousche and Normann, 1998; Williams et al., 1999a,b; Rousche et al., 2001; Csicsvari et al., 2003; Kipke et al., 2003; Nicolelis et al., 2003; Vetter et al., 2004; Patrick et al., 2006; Jackson and Fetz, 2007; Musallam et al., 2007; Kozai et al., 2012a; Kim et al., 2013). There is a general consensus in the microelectrode electrophysiology community that there is a lack of standardization in surgical methodology, insertion technique, and electrode manufacturing. Systematic and quantitative study of each of these array types is necessary to prove their benefits and weaknesses in the context of electrode performance and failure.

A timeline of important phases occurring during an electrode's implant lifetime is presented here to guide the implant durations for the study of electrode longevity. In general, the key phases known to be associated with chronic electrode performance used are as follows: acute (Ward et al., 2009), recovery (Turner et al., 1999; Winslow and Tresco, 2010), chronic (Williams et al., 1999a; Polikov et

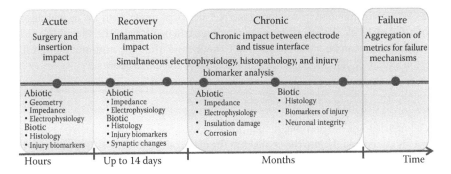

Acute	Recovery	Chronic	Failure
Surgery and insertion impact	Inflammation impact	Chronic impact between electrode and tissue interface Simultaneous electrophysiology, histopathology, and injury biomarker analysis	Aggregation of metrics for failure mechanisms

Abiotic
• Geometry
• Impedance
• Electrophysiology
Biotic
• Histology
• Injury biomarkers

Abiotic
• Impedance
• Electrophysiology
Biotic
• Histology
• Injury biomarkers
• Synaptic changes

Abiotic
• Impedance
• Electrophysiology
• Insulation damage
• Corrosion

Biotic
• Histology
• Biomarkers of injury
• Neuronal integrity

| Hours | Up to 14 days | Months | Time |

Figure 5.1 A timeline of major events after electrode implantation served as an aid to determine implant duration for each animal.

al., 2005; Winslow and Tresco, 2010), and failure (Figure 5.1). The "acute" phase is described by the implantation pathological responses that occur due to the mechanical trauma induced in the tissue due to electrode insertion. The "recovery" phase involves the initial electrode exposure to biological tissue and histopathological changes due to initial foreign body response. The "chronic" phase is affected by the long-term structural changes that occur on the recording surfaces of the electrodes combined with the chronic neuroinflammatory response. It is known that not all electrodes fail at the same time within an implanted array and that there can be substantial variation in the lifetime. Figure 5.1 identifies these phases and provides a list of the individual metrics typically studied during each phase.

5.2 Morphological properties

5.2.1 Scanning electron microscopy imaging

To evaluate the structural changes in chronically implanted microelectrode arrays, all arrays should be imaged before implantation and after explantation using either optical techniques or scanning electron microscopy (SEM) such as a variable pressure scanning electron microscope (Hitachi S-3000N VP-SEM). For SEM approaches, the environmental mode can be chosen to enable the direct placement of the samples into the SEM chamber without the use of carbon or conductive coating. This is very critical especially for pristine electrodes as they will be implanted into an animal after the SEM imaging procedure. Such a method is also used to cause no damage to the arrays from the imaging process. When the preimplant and postimplant images are taken, the electrode arrays should be handled with care and microtools to prevent any damage to the electrode shanks. The following parameters are commonly used to take pre- and postimplant images: environmental secondary electron detector mode with an acceleration potential of 12 kV. The working distance range can be set between 18 and 40 mm. Magnification can be varied according to need, and often the minimum magnification can be set at 20× depending on the size of the electrodes.

5.2.2 Imaging observations

A comparison of the preimplant and postexplant SEM images of electrode arrays provides deep insight into the changes at the electrode recording surface morphology. Images before implantation allow one to observe any preimplant defects in the electrodes due to the manufacturing process. For example, several defects occurring as a result of manufacturing most commonly present in the recording surfaces and the interface between the recording surface and the insulation. The manufacturing defects mostly commonly present include cracks in the insulation material, nonuniform insulation, bent and broken recording tips, cracks on the recording tips, and nonuniform deinsulation (Figure 5.2). Figure 5.2 shows examples of microwires with preimplant manufacturing imperfections for four animals (3–6 months implant period). These images were taken before electrode implants. Note that not all the electrodes within an array have issues, but there is a large

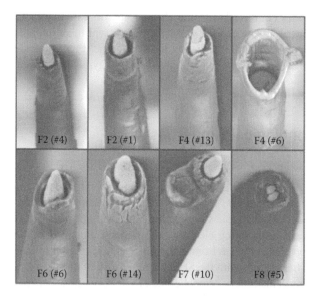

Figure 5.2 Manufacturing variability. SEM images show large variation between electrodes within the same Pt/Ir array. Shown here are eight wires from four arrays with imperfections as a result of manufacturing process. One of the wires is missing the recording tip, whereas others have cracks in the insulation/recording tip and uneven insulation at the interface of the recording surface/insulation.

variation between electrodes within an array. Preimplant and postexplant SEM images allow for a qualitative comparison of the changes that occur at the electrode recording surface and insulation material. Figure 5.3a shows such changes occurring after chronic implantation in animals where the electrodes underwent moderate to minimal changes at the recording surfaces. Here it was observed that Pt/Ir undergoes little corrosion for implants up to 6 months, as would be expected for an inert metal. However, there is a great variability among individual electrodes even within the same array. Although electrode corrosion does not appear to be a significant problem with Pt/Ir, there is severe delamination and cracking of insulation material. Gross morphological changes can occur at the recording surface in the form of bent tips, cracks appearing in the insulation material, and insulation delamination. Figure 5.3b shows 12 representative electrodes from seven long-term animals (3–6 months) where severe damage to the insulation material can be observed. Electrodes that undergo gross morphological changes usually have poor electrode performances during the chronic lifetime. Insulation

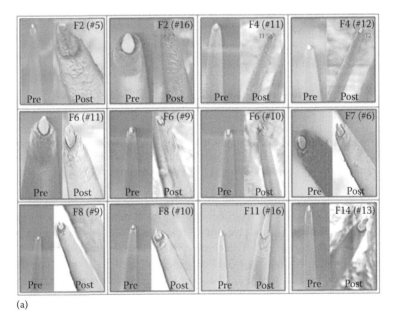

(a)

Figure 5.3 (a) Preimplant and postexplant SEM. Preimplant and postexplant SEM images show minimal to moderate changes at the recording site structure. However, recording site changes occur as a result of corrosion of the recording surface for long-term implants. Shown here are three 6-month implants (F2, F4, and F6), two 3-month animals (F7 and F8), one 15-day animal (F11), and one 7-day animal (F14), respectively. A crack at the recording site can be observed to be developing in the electrode F2 (#5). *(Continued)*

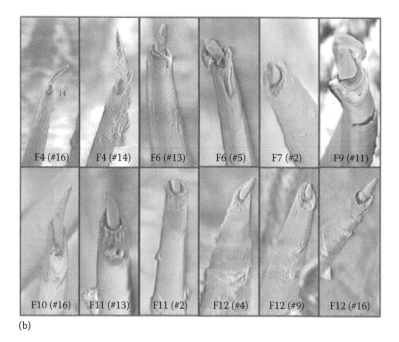

| F4 (#16) | F4 (#14) | F6 (#13) | F6 (#5) | F7 (#2) | F9 (#11) |
| F10 (#16) | F11 (#13) | F11 (#2) | F12 (#4) | F12 (#9) | F12 (#16) |

(b)

Figure 5.3 (Continued) (b) Insulation deterioration postexplant SEM. Postexplant SEM of individual microwire to indicate deterioration in electrode insulation for parylene-C-coated Pt/Ir microwires. The deterioration occurs in the form of delamination and cracks. Although the insulation deterioration varies among microwires even with the same array, we observed it to be present in all the wires across animals for all implant durations (7 days–6 months).

damage and delamination can result in a decrease in electrode impedance during the chronic lifetime and affected functional performance (Figure 5.4). Figure 5.4 shows such an example for four electrodes from three long-term animals (F4 and F6: 6 months; F9: 71 days) where electrode impedance declined during the latter part of the implant period. As would be expected, as the electrode is undergoing these morphological changes, the functional performance (electrode yield) becomes very poor.

5.3 Electrical properties

5.3.1 Impedance testing procedure

Impedance measurements can be performed in vivo using a NanoZ impedance tester (TDT, FL) or similar equipment (Prasad and Sanchez, 2012). Briefly, the NanoZ measures impedance by applying a small sinusoidal voltage of 4 mVpp at

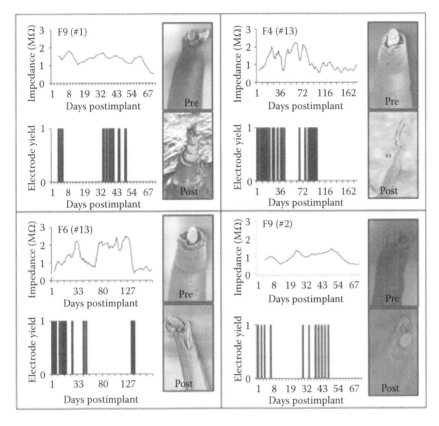

Figure 5.4 Effect of insulation damage: preimplant and postexplant SEM images are shown for two Pt/Ir electrodes that undergo delamination during the 6-month implant duration (F6 and F9). Insulation delamination results in a decline in 1 kHz electrode impedance combined with poor functional performance (low yield).

specific frequencies of 1 Hz, 2 Hz, 5 Hz, 10 Hz, 20 Hz, 50 Hz, 100 Hz, 200 Hz, 500 Hz, 1 kHz, and 2 kHz, respectively. The NanoZ impedance meter uses a voltage divider circuit shown in Figure 5.5 to calculate the electrode impedances. The circuit can be solved for voltages according to the Ohm's law. The NanoZ measures the impedance by applying a known voltage V1 and measuring V2. The equation for Ohm's law can then be solved to obtain Zx (electrode impedance). The NanoZ applies a constant voltage of 4 mVpp sinusoidal waveform for V1, which causes a maximum test current of 1.4 nA root-mean-square through Zx when Zx approaches zero. To minimize errors, it is best to repeat the impedance measurement process multiple times (20–50 is appropriate) and take an average value. Most impedance measurement equipment will have a built-in function to perform this operation.

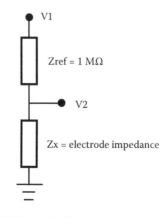

$V1/V2 = 1 + Zref/Zx$

Figure 5.5 Simplified impedance measurement circuit used by the NanoZ impedance tester to calculate electrode impedance.

Because of the large difference in impedance between implanted microelectrodes relative to their distant low impedance reference electrode, the method presented here is suitable for impedance measurements *in vivo* (Brett and Brett, 1993; Williams et al., 2007). Impedance is calculated using a ratio of the measured voltage (V2) over current. The measured voltage comprises of the overpotentials at two electrode interfaces (working and the reference electrode) and the voltage drop (product of the current and the tissue medium resistance) (Merrill et al., 2005). The overpotential itself occurs as a result of three mechanisms: the ohmic, the concentration, and the activation overpotentials (Webster, 1997). It is difficult to tease out one component from the other in *in vivo* measurements such as described previously. Therefore, the impedance reported here is calculated from the observed voltage (V2), which occurs as a result of all the three components: the two interfacial overpotentials and the voltage drop in the tissue medium.

To obtain baseline measurements from electrodes, impedance spectroscopy can be performed *in vitro* before implantation in 0.9% phosphate-buffered saline (PBS). Impedance in this preparation is measured with respect to a low impedance stainless steel reference wire dipped in the same solution. Again, the two-electrode method is suitable for measuring microelectrode impedance owing to the large difference in impedance between the electrode being tested and the distant low impedance reference wire (Brett and Brett, 1993; Geddes, 1997; Williams et al., 2007).

5.3.2 Recording procedure

Comprehensive understanding of electrode performance involves coupling impedance measurements with *in vivo* neural recording. Test beds for rapidly evaluating electrode arrays can be constructed to monitor multiple animals simultaneously in awake behaving conditions, as shown in Figure 5.6. Animals can be recorded 7 days a week with impedance measurements for all electrodes occurring every recording session. In this example setup, a headstage zero insertion force (ZIF) connector with a unity gain amplifier is used to interface the implanted MEA with a preamplifier (TDT, FL), which is connected to a real-time signal-processing system (RZ2, TDT, FL). A custom-made software program (TDT, FL) is commonly used to record neural activity at 24414.06 Hz from the electrode array. To obtain single neuron recordings, neural signals are band-pass filtered between 0.5 and 6 kHz, and online spike sorting (Lewicki, 1998) based on thresholding the signals can be used to isolate individual neuronal activity around the microelectrode tips. Offline spike sorting should be performed to verify isolated units and clean the signal to get rid of artifacts not removed during online sorting. Waveforms are classified as single units based on the repeatability of waveform shape and peak-to-peak amplitude with respect to the background noise (Suner et al., 2005). Electrophysiological recordings are quantified via array yield, which is defined as the percentage of electrodes within an array that are able to isolate at least one neuron during a recording session. An electrode is said to isolate a neuron (single units) when at least 100 reproducible biphasic waveforms can be collected during a session.

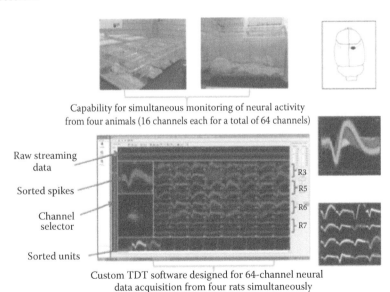

Figure 5.6 Experimental setup allowing simultaneous monitoring of neural signals from four animals. The screenshot shows the custom designed software for neural data acquisition. The top picture on the right shows the implant location. The bottom picture shows example units recorded during a session.

89

5.3.3 Impedance and electrophysiological observations

There are identifiable trends between daily impedance measurements and the array yield observed in chronic neural implants. Typically, high array yield is correlated with low impedance values (<200 kΩ) during the recovery period after implant. As the impedance values tend to increase during the recovery period and the beginning of the chronic phase, the array yield tended to decrease. The array yields (a) and the impedance measurements (b) in eight animals for implant periods greater than 6 weeks are shown in Figure 5.7. The figure shows that there is sharp drop in array yield after the second week when there is almost a twofold increase in impedance. A slight increase in yield is also observed for chronic animals during 6 weeks of implant duration, where the impedance reduces for all the animals. However, the array yield decreases if the impedance values decrease too much by falling lower than the range (40–150 kΩ).

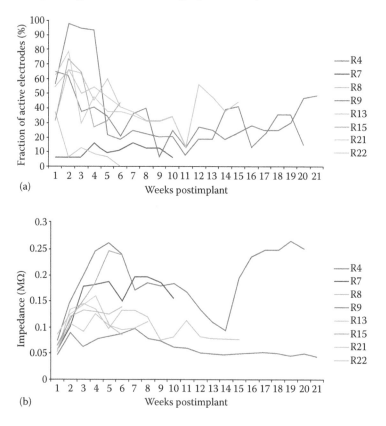

Figure 5.7 Aggregated data from 8 chronic animals to show the trends between the weekly array yield (a) and the impedance values (b) for long-term implants ($\geq$6 weeks).

90

5.3.4 Probabilistic model for unit detection

On the basis of the observations described in the previous section, a model can be designed to describe the probability of detection of a single unit given a particular impedance range. This relationship is powerful for predicting the expected performance of an electrode from electrical measurements and is shown in Figure 5.8. To construct the plot, the observed impedance range can be divided into 10 kΩ bins up to 400 kΩ for the total number of units from each electrode, given a particular impedance range during a recording session to be counted. To build the model, data should be aggregated from all sessions from animals. The probability of unit detections (purple dots) shown in Figure 5.8 is calculated by normalizing the total number of units for each of the impedance range by the total number of sessions and animals for the training dataset. The largest peaks can be observed

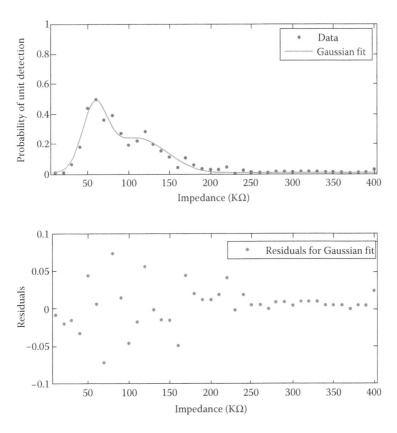

Figure 5.8 Top plot shows a Gaussian model fitted to the probability of unit detections obtained from 13 chronic animals for the impedance range of 0–400 kΩ. The bottom plot shows the residuals obtained after fitting the model to the data.

91

to be in the 40- to 150-kΩ range where most of the units are detected. This peak in Figure 5.8 suggests that electrode impedance can be used as one of the predictors of electrode performance and failure.

To predict electrode yield, a Gaussian model can be used to model the general probability trends for different impedance ranges. A Gaussian model is selected because the shape of the probability curve appears to be Gaussian. The main criterion while selecting the model is to take the minimum of the sum of squares of errors (SSE) and at the same time maximize determination of the coefficient R^2 value. The Matlab Curve Fitting Toolbox can be used to perform this analysis. Given this relationship, it is possible to input new impedance measurements from future animals and recording sessions and obtain an estimate of the expected array yield. For these data, the probability is described by the following general Gauss2 model equation, where y is the probability of unit detection and x is the impedance value:

$$y = a1*\exp(-((x - b1)/c1)^2) + a2*\exp(-((x - b2)/c2)^2).$$

Coefficients (with 95% confidence bounds) are as follows:

$a1 = 0.3939 \ (0.2813, 0.5066)$
$b1 = 5.893 \ (5.66, 6.127)$
$c1 = 2.101 \ (1.57, 2.632)$
$a2 = 0.2377 \ (0.2036, 0.2717)$
$b2 = 10.96 \ (9.323, 12.6)$
$c2 = 5.206 \ (3.457, 6.956)$

Goodness of fit data are as follows:

$SSE = 0.02894$
$R^2 = 0.9601$

A nominal threshold value of 30% of the maximum peak in Figure 5.8 is used in a binary classifier model to determine whether a given probability corresponds to a unit or not from each electrode. The model predicts the occurrence of a single unit given an impedance value. The prediction accuracy of the model is calculated by comparing the number of times the model correctly predicted the presence of a unit from each electrode within a session given the impedance of that electrode. The accuracy is then averaged over all the recording sessions from all the animals in the test set to obtain a mean prediction accuracy of 81.92% ± 19.7%.

5.4 Tissue properties

Histological analyses of microelectrode arrays focus primarily on in-depth assessments of microglial reactivity after electrode implantation because microglia are cells rapidly responding to CNS tissue perturbations and are therefore the most sensitive sensors of brain pathology. Four common antibodies are used to illuminate various functional aspects of microglial responses and microglial subpopulations: (1) Iba1 as a robust marker for visualizing all microglial cells present in a given section; (2) ED1 staining against the CD68 antigen to identify active macrophages (Dijkstra et al., 1985), but not necessarily all activated microglia because many activated cells are not engaged in phagocytosis and are thus ED1-negative (Graeber et al., 1998); (3) antiferritin staining to identify those microglia involved in the sequestration of free iron, which may leak into the CNS as a result of BBB compromise; (4) OX-6 immunostaining to identify antigen presenting cells, which typically express the MHC class II antigens recognized by OX-6 antibody. MHC antigens can be expressed by microglia or blood-borne immune cells.

5.4.1 Histopathology

The tissue reactions related to electrode implantation are investigated using fluorescent immunohistochemistry. To process the tissue, the head of the subject is rinsed in several changes of 0.1 M PBS (pH 7.4) for 2 to 3 hours. The brain tissue is accessed by removing the top of the skull and the electrode cap as one assembly. Before cryosectioning, tissues are cryoprotected using 30% sucrose in PBS. The brain is embedded with tissue freezing medium (Cat# H-TFM, Triangle Biomedical Sciences, Durham, NC). The cerebrum is embedded for sectioning in the horizontal plane. Frozen sections are cut on a cryostat at 20 μm thickness.

5.4.2 Immunohistochemical procedures

Microglial cells in brain sections are visualized immunohistochemically using the rabbit polyclonal primary antibody Iba1, directed against the ionized calcium binding adaptor molecule (Wako, 019-19741, diluted at 1:800). This antibody binds to all microglial cells irrespective of their activation or degeneration state. Phagocytic microglia are identified by labeling with antibody ED1 (mouse antirat monocytes/macrophages [CD68], Chemicon; MAB1435, MAb; IgG; 1:300). Immunolabeling for the iron storage protein, ferritin, is performed using antiferritin polyclonal antibody (rabbit antihorse spleen ferritin, Sigma, F6136, diluted at 1:800). Ferritin is an iron storage protein that becomes expressed in some microglia, which in human brain are often dystrophic rather than activated. Ferritin staining of microglia seems to be induced under certain injury/disease conditions when the blood–brain barrier is disturbed and there is a need for iron sequestration. The ferritin antibody also binds to oligodendrocytes in the normal uninjured central nervous system (CNS). The intensity of immunological activity is assessed by OX-6 expression (mouse antirat MHC Class II RT1B, IgG1, AbD SeroTec, MCA46G, diluted at 1:400), which recognizes antigen presenting cells. The OX-6 antigen expression is increased after injury during microglial activation and is used as a generic measure of the intensity of the neuroinflammatory response.

For fluorescent immunolabeling, sections are blocked with blocking buffer (10% goat serum in PBS) and incubated with the primary antibody followed by several washes and incubation with appropriate secondary fluorescent antibodies. For double labeling of different antigens in the same section, a combination of primary antibodies is applied to sections as cocktail primary antibodies. Cocktail secondary antibodies are then used for visualizing binding sites of primary antibodies. Monoclonal primary antibodies are visualized using goat antimouse IgG conjugated with Alexa Fluor 488 (Molecular Probes A11001; Eugene, OR; used at a dilution of 1:400). Polyclonal primary antibodies are visualized using goat antirabbit IgG conjugated with Alexa Fluor 568 (Molecular Probes A11011; diluted at 1:400). Sections are then coverslipped with GEL/MOUNT (M01, Biomeda Corp., Foster City, CA).

5.4.3 Observation and imaging

For the images shown here, slides are examined with a Zeiss Axioskop 2 microscope. Digital images are captured with a Spot Slider 3 digital camera (Diagnostic Instruments Inc., Sterling Heights, MI). In double-immunofluorescent-labeled preparations, micrographs are prepared using images pseudo-colored or digitally merged from images captured at single fluorochrome using Adobe Photoshop software (Adobe Systems Inc.; San Jose, CA).

5.4.4 Quantitative morphometry

The quantitative analysis of antigen expression on sections is conducted on images using Image J program (NIH, Bethesda, MD). Black-and-white images are captured with a 20× objective. The total positive area of an antigen expression per unit section area (i.e., per microscopic field ×20; total area size 267320.12 μm^2) is scored in an image captured from each side of brain by using Image J program. The mean values and standard deviation of the counts are computed. An intact microglial cell is defined as a cell with a cell body area larger than 30 μm^2 and less than 300 μm^2 for Iba1, ferritin, and OX-6 labeling. The ED1-positive particle is defined as areas from 1 to 10 μm^2. The difference between the electrode implanted side and the contralateral side is interpreted to reflect an approximation of microglial cell function.

5.4.5 Histopathologic analysis

Almost all chronic microelectrode implants are marked by strong increases in the four markers regardless of survival times. However, animals with the longest survival times (R4, 9, 13 shown here) revealed relatively modest upregulation in overall microglial reactivity (Iba1 staining) compared to shorter surviving rats, suggesting that over time the neuroinflammatory response may be declining (Figure 5.9a).

94

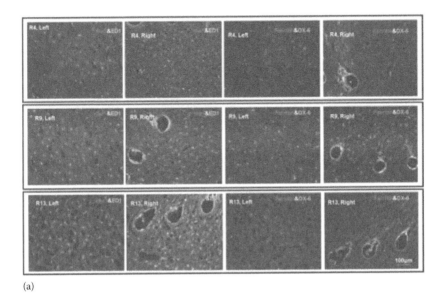

(a)

Figure 5.9 (a) Microglial reactivity shown for the three longest-term animals (R4: 260 days; R9: 217 days; and R13: 187 days) showed modest upregulation in overall microglial reactivity (Iba1 staining), suggesting that neuroinflammatory response may be declining for the longest-duration animals.   (*Continued*)

R7 is an animal in which significant intraparenchymal bleeding was noted during surgery, and it also revealed the presence of microglial degeneration during microscopic examination, which was generally uncommon during the recovery and chronic survival periods. In animals with bleeding issues, all showed robust increases in ferritin expression (Figure 5.9b), suggesting that there was a substantial need for the sequestration of free iron that may have leaked into the parenchyma. Incidentally, the array yield in these animals was generally quite poor.

A remarkable and novel observation concerns the very consistent presence of dystrophic microglial cells in animals from the acute phase cohort (Figure 5.9c). Dystrophic microglia have been known to become apparent primarily in aged human brain where they are likely to reflect microglial senescence (Streit et al., 2009). Quite strikingly though, the dystrophic microglia observed here early after electrode implantation reveal the same characteristic morphological phenotype seen in aged humans, i.e., the appearance of cytoplasmic fragmentation (cytorrhexis). Only some of the longer surviving animals show evidence of microglial cytorrhexis, which often coincided with poor electrode performance (Figure 5.9b), but overall the relationship between the presence of microglial dystrophy and the poor array yield is inconsistent. Some animals with moderate or good electrode

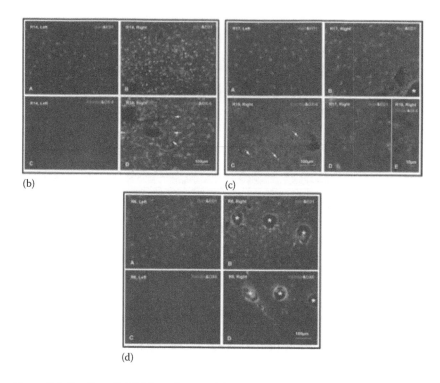

(b) (c)

(d)

Figure 5.9 (Continued) (b) Microglial reactivity is pronounced 7 days after implantation in an animal with modest electrode performance (R14). (a and b) Double immuno-fluorescent staining with Iba1 and ED1 antibodies reveals the presence of normal ramified microglia on the unimplanted (left) side of the brain and many activated and phagocytic microglia near the implanted device (electrode tracks not in field). Microglia can be identified as phagocytic by intracellular colabeling with ED1; most cells in the field are ED1 positive. (c and d) Pronounced upregulation of ferritin immunoreactivity in many activated/phagocytic microglia on the implanted (right) relative to the control (left) side. An occasional ferritin+ cell (arrow) shows cytorrhexis (arrow). Scale bar applies to A–D.

performance do indeed reveal a benign surgical and histopathological picture in that there is no intraparenchymal bleeding, little encapsulation of electrode tracks, low numbers of activated microglia, and low ferritin expression (Figure 5.9d).

5.4.6 Cerebrospinal fluid and blood collection

To evaluate chronic biomarkers of injury throughout the duration of implant, cere-brospinal fluid (CSF) can be collected from the cisterna magna (fourth ventricle) (Pegg et al., 2010). This analysis focuses on pNF-H, reported as a novel biomarker for axonal injury, and a heavily phosphorylated axonal form of the neurofilament subunit NF-H. Ongoing axonal damage or degeneration causes the release of this biomarker into the blood and CSF, otherwise not detectable in uninjured control animals. To collect this biomarker from CSF, anesthetized animals are placed in a stereotaxic frame, and a small incision (~0.5 inch) is made between the animal's ears below the caudal ridge of the skull. The muscle layers are separated gently to reach the cisterna magna, which is a small region between the base of the skull and where the spinal cord begins. The dura is exposed at this region and CSF can be acquired by puncturing the dura without the need to perform laminectomy. A 28-gauge needle attached to a microinjector (Stoelting, Wood Dale, IL) on a stereotaxic arm is used to puncture the dura and slowly withdraw approximately 100 μL of clear CSF. The animal's skin is sutured back on withdrawal. The CSF is then put over ice for 20 minutes and transferred to an Eppendorf tube and centrifuged at high speed for 10 minutes. The clear CSF is transferred to another Eppendorf tube and stored in a −80°C freezer for biochemical analysis. Care should be taken during CSF withdrawal so that it is never contaminated with blood. Blood is collected from animals by tail vein puncture. To do so, animals are anesthetized using isoflurane, and their tail vein is punctured using an 18-gauge needle. The blood is allowed to cool over ice for 20 minutes and centrifuged at maximum speed for 10 minutes. The clear supernatant is separated from the pelleted blood cells on centrifuging and transferred to an Eppendorf tube before storing in a −80°C freezer.

5.4.7 Biochemical analysis

The blood and CSF samples enables one to continuously monitor the pNF-H biomarker for providing insights into the CNS injury caused due to the electrode implants. Sustained elevation indicates continuing damage in the microenvironment around the implanted probes. All measurements are normalized with respect to control rats with no injury which does not express measurable levels of pNF-H in serum or CSF. A summary of pNF-H levels expressed in blood and CSF from representative animals for acute, recovery, and chronic phases is shown in Figure 5.10. All acute animals had the least levels of pNF-H (<0.1 ng/mL) present in both the blood and the CSF due to the electrode implantation in the neural tissue. Although the biomarker is present, the levels are not significantly high, suggesting minimal trauma induced during surgery. pNF-H is present in animals in the recovery period, and its levels decrease for R10 during the 16-day implant period. The pNF-H levels for R10 after surgery are significantly higher (~0.8 ng/mL) than most of the animals, although the levels slightly reduce (0.6 ng/mL) on day 6 and significantly reduce (<0.2 ng/mL) by day 16, suggesting a spike in pNF-H due to the surgery, which improves as the animal recovers from the implant trauma. Figure 5.10 shows another recovery animal (R16), where pNF-H levels are low and similar to other animals at samples collected on day 7. The data from chronic

97

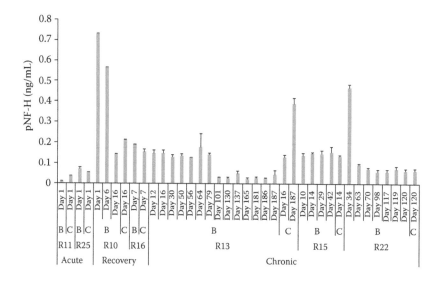

Figure 5.10 Biochemical analysis for the presence of pNF-H in blood (B) and CSF (C) collected at regular intervals for representative animals in each of the three phases (acute, recovery, and chronic).

animals show elevated levels of pNF-H for long periods of time before reducing at the time point 60+ days, depending on the animal. For example, pNF-H levels for R13 remain elevated until day 79 after which they reduced by approximately threefold and thereafter remain at those levels until the animal is euthanized on day 187. The levels never reduce to zero, suggesting the presence of these electrode arrays continuously causes axonal injury. There are also periods of times when the levels spike up but generally reduce afterward. Both CSF and blood had similar levels of pNF-H on the same sampling days for all animals most of the times, indicating that either can be used for long-term analysis.

5.5 Holistic abiotic and biotic analysis

A holistic analysis of the biotic, abiotic, and functional metrics is presented in Figure 5.11. Here, it is possible to simultaneously determine how the interrelationships between impedance and biomarkers of injury affect the functional performance of the array throughout the entire duration of implantation. Among the 11 animals presented in this figure, several trends emerge. In general, the animals that have the least axonal injury and impedance in the range of 40 to 150 kΩ produce the best performance (70–100% yield). Animals that have the poorest performance (0–25% yield) (R4, R5, R7, and R8) produce at least one failure in either the biotic or the abiotic metrics. For R4 and R7, both the impedance is too

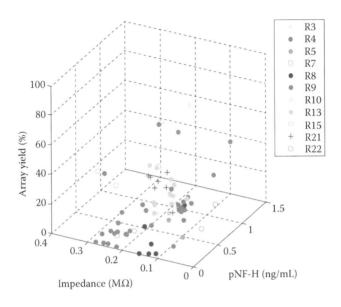

Figure 5.11 Coupling of impedance measurements, electrophysiological yield, and biomarker analysis to determine how the interrelationships between impedance and biomarkers of injury affect the functional performance of the electrode array during the implant duration. Animals with poorest functional performance (R4, R5, R7, and R8) indicated at least one failure in either abiotic or biotic mode with either too high electrode impedance or high axonal injury biomarker (pNF-H) levels. Animals with good functional performance (moderate-to-good electrode array yield in R3, R9, R10, R13, and R15) during their respective implant duration had time points where they had optimal electrode impedance (40–150 kΩ) and low levels of pNF-H.

high, and there are high levels of axonal injury indicated by the pNF-H. R8 has only one failure with high pNF-H; however, the impedance is in the appropriate range. On the other end of the spectrum, the best performing animals (R3, R9, R10, R13, and R15) have time points where there is optimal impedance and low axonal injury. The highest locations in the array yield axis indicate these time points. R9, R10, R13, and R15 are not perfect, although and there are time points where the biotic and abiotic metrics were suboptimal and moderate performance (25–70% yield) is obtained. These form a cluster of points and also contain animals R13, R21, and R22. In addition to the general trend in how the metrics affect the performance levels, the variance in the cluster of points for each animal illustrates the time-varying nature of the interactions. All of the samples are not tightly clustered over all time. There are periods where each of metrics plays a differential role in the overall failure or performance of the array. We have shown through this chapter that each of the previously discussed metrics show some

99

power for predicting electrode performance. The tissue–electrode interface is a complex environment where electrode performance is likely governed by interplay of both biotic and abiotic interactions.

5.6 Conclusion

This chapter presents a study of the *in vivo* abiotic and biotic characterization of microelectrode arrays in chronic implants. The approach combined SEM imaging, impedance spectroscopy, and histopathology with electrode functional performance so that better understanding of long-term microelectrode performance can be achieved. Abiotic and biotic measurements were performed over the implant duration, revealing changes that occurred at the electrode–tissue interface affecting the functional performance of the microelectrodes. Preimplant SEM imaging showed significant variation between electrodes even within the same array indicating manufacturing imperfections. Postexplant SEM imaging indicated that electrodes exhibited deformation of the recording sites such as bent or cracked tips. In addition, deterioration of insulation material was observed among all electrodes in all animals in the form of insulation delamination and cracking. Generally, electrode impedance reduced chronically in electrodes that had insulation deterioration and poor functional performance. Histopathological evaluation suggested modest elevation of microglial markers for longest-term animals. In contrast, in this study, it is shown that insulation deterioration, manufacturing, and corrosion were the likely abiotic factors contributing to the electrode failure. In addition to the chronic abiotic and biotic effects, evidence is provided for vascular disruption, and concurrent intraparenchymal hemorrhaging during brain implantation is a major biotic factor that likely contributes to electrode failure.

The results in this chapter indicate a need for precision and consistency of manufacturing of microelectrode arrays, an improvement that will likely play a major role in the long-term performance (Prasad and Sanchez, 2012; Prasad et al., 2012; Streit et al., 2012). Of particular interest is the recording site itself and interface between that site and the surrounding insulation. Preimplant variations provide a weak or fault site on the microelectrodes that will lead to an accelerated deterioration of the recording surface of individual electrodes. Electrodes with imperfections before implantation are more susceptible to morphological changes during the implant period. Manufacturing defects at the recording sites also provide fault sites where accelerated deterioration can happen, resulting in a decreased electrode impedance and poor functional performance. Therefore, to mitigate the variability arising out of the manufacturing process, greater quality control measures should be employed preimplant to ensure that implanted electrodes are all of good quality. The manufacturer of the electrodes typically manipulates the amount of deinsulation for each electrode during fabrication to control the impedance. Although the intent of the deinsulation process is to control impedance, observations from the electrodes reveal that the process is not as precise as would be desired and there is inconsistency in the manufacturing process. Variance in the starting impedance and manufacturing is compounded by changes in impedance after surgery.

In addition to the electrode insulation interface, results strongly suggest the bulk insulation itself is an important contributor leading to electrode failure. Insulation delamination and cracking is evident on all array types. Electrodes that have insulation damage preimplant are more prone to undergo insulation deterioration during the chronic implant. Insulation damage results in leakage resistance and parasitic capacitance (Kane et al., 2013) and continuous decrease in impedance, which leads to an accelerated damage causing decrease in electrode functional performance. These results indicate that electrode delamination and cracking are significant failure modes and steps must be taken to evaluate better electrode insulation materials in future.

Material differences should also be considered while choosing electrode metals for recording purposes. In previous studies, it has been observed that tungsten undergoes rapid corrosion *in vitro* (Patrick et al., 2011) and *in vivo* for implants as early as 1 week after surgery (Prasad et al., 2012). The deterioration of tungsten continued in the chronic phase in the form of recording sites becoming more recessed. Therefore, the choice of recording material is one of the contributing abiotic failure modes for tungsten electrodes. However, Pt/Ir provides better recording surfaces and corrosion and is not such a big issue as with tungsten.

One of the experimental design factors that enabled deeper insight into electrode failure is the coupling of electrode array functional performance with impedance spectroscopy to study if there was a functional relationship between them. Electrode array functional performance is quantified by array yield, which was defined as the percentage of electrodes in an array that are able to isolate single units. Higher array yields are desirable for a neuroprosthetic application. In general, for microelectrodes, 1-kHz impedance magnitude trends were similar for long-term subjects where impedance increased progressively in the first 2 to 3 weeks and then decreased in the subsequent weeks with occasional increases and decreases in the chronic period. Array yield generally declines during this initial 2- to 3-week period after implant, attributed as a result of surgical trauma in other studies (Biran et al., 2005). Poor array yield (20–40% yield) is observed when impedance was high (>1.5 MΩ), and yield increases when impedance is low during the late chronic period (>12 weeks) for long-term animals. There exists an inverse relationship between the array yield and the electrode impedance at 1 kHz for all long-term animals.

There is a cascade of biological reactions that occur after an electrode implant surgery that in brief includes activation and recruitment of astrocytes, microglia, and macrophages at the implant site to isolate the implanted device and to repair the damaged tissue (Polikov et al., 2005; Thelin et al., 2011; Kozai et al., 2012b). Several studies have pointed out the time course of such biological reactions occurring at the tissue–electrode interface that begin as early as right after insertion and continue as long as the implant is in the neural tissue (Szarowski et al., 2003; Freire et al., 2011; Prasad et al., 2012). Of course, the most obvious and inevitable biological consequence of intracerebral electrode insertion, or for that matter any type of traumatic brain injury, is a neuroinflammatory response. Neuroinflammation is mediated primarily by activated microglial cells. Electrode implantation elicits widespread microglial activation at the implantation site, which in most cases becomes somewhat attenuated over time. However, microglial degeneration, evident as cytoplasmic fragmentation, is a direct consequence of oxidative stress brought about by influx of free iron through intracerebral bleeding. Microglial degeneration can occur during early or late time points, especially in animals where there was significant intracerebral bleeding, thus supporting the concept that microglial cytoplasmic fragmentation occurs as a direct consequence of iron-mediated oxidative stress.

These results serve to further illuminate the problem of electrode failure by focusing attention on two major failure modes: abiotic and biotic. Future efforts geared toward prolonging the functional lifetime of electrodes should thus be focused on further improving electrode manufacturing practices and on innovative ways of minimizing intracerebral bleeding and associated oxidative stress.

Exercises

1. The assessment of microelectrode failure should be studied from which of the following perspectives?
   a. Biotic analysis
   b. Abiotic analysis
   c. Functional analysis
   d. All of the above
2. What are the major types of cells involved in neural tissue response to chronically implanted microelectrode arrays? Describe how those cells can be affected and can affect the abiotic components of microelectrode recordings.

102

Chapter 6 Neural decoding

But the truth is, you have to be willing to subject yourself to failure, to be bad at something, to fall on your butt and do it again, and try stuff you've never done. That's the ideal mind-set in sports and in life—you have to be willing to have people laugh at you at first. Why? Because you need to keep challenging yourself. That's the whole idea behind seeking out things you're not good at: It forces your mind to engage.

Laird Hamilton
Force of Nature (2010, 43)

Learning objectives

- Identify neural features used as control signals for neuroprosthetics.
- Understand the use of feedback and errors in building neuroprosthetic decoders.
- Learn how to design and deploy dynamic decoder systems that enable customized configurations, on-the-fly adaptation, and on-demand provision of real-time performance.

6.1 Introduction

Neuroprosthetics have tremendous potential to improve the life of patients with impaired nervous systems and to introduce new paradigms of human–machine interaction that can impact many aspects of daily life. An integral part of the design of neuroprosthetics includes the development of decoders and encoders that translate thoughts (neural activity) into machine-readable signals that can be used to perform a functional outcome. The development of neuroprosthetic decoders is multidisciplinary because it synergizes theory and experimentation in computational neuroscience, electrophysiology, adaptive signal processing, controls, machine learning, and cognitive architectures. The main goal of the decoder design is to enable closed-loop, direct communication, and control among the brain, computers, and devices. The ability of the decoder to deliver and integrate the net benefits of computational networks, multiscale brain signaling, and sensory stimuli depends on the design of a control architecture that has the capability to constantly generalize to the ever-changing conditions between the neuroprosthetic user and the neuroprosthetic device during task learning. These changes include: (1) the modifications of brain activity that result from response to novel stimuli, brain injury, and new task acquisition; (2) the dynamics in environmental conditions; and (3) the ongoing translation of the user's goals to situational context. Because neuroprosthetics provide novel inputs and outputs for the nervous system, which are augmentative to what is provided by natural biology, and both the user and the engineered system adapt to each other, the user and the computer are in a symbiotic relationship while interacting with a complex, dynamic environment. To create neural interfaces that are capable of integrating all of these aspects and facilitate performance, the designer should build a framework using the biologic perception action reward cycle (PARC) that learns during neuroprosthetic use and divides the "intelligence" between the user and the neural decoder. This behavior operates on a series of events or elemental procedures that promote specific brain or behavioral syntax, feedback, and repetition over time (Calvin, 1990). The sequential evaluative process of interacting with the environment is always ongoing and adheres to strict timing and cooperative–competitive processes. Figure 6.1 shows the basic components of sensorimotor systems engaged with the environment. With sensory, motor, and reward processes, intelligent sensorimotor goal-directed control can be built with closed-loop

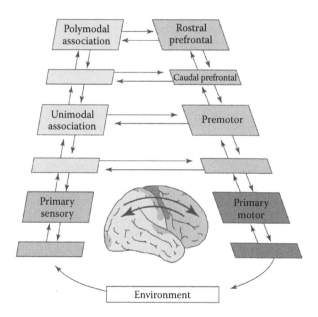

Figure 6.1 The major components of the perception (left)–action (right) cycle. (Reprinted with permission from Fuster, J. M., Upper processing stages of the perception-action cycle, *Trends Cogn. Sci.*, 8 (4), 143–5, 2004. doi: 10.1016/j.tics.2004.02.004.)

104

mechanisms, which continuously adapt internal and external antecedents of the world, express behavior in the environment, and evaluate the consequences of those behaviors to promote learning. Collectively, these components contribute to organizing behavior in the nervous system (Fuster, 2004). The entire process is regulated by external environmental and internal neurofeedback, which is used to guide the adaptation of computation and behavior. The sophistication of the control models developed in this chapter is such that they must interact with the user and implement action evaluations and informed decisions in real time (within the 100 ms of the user's PARC) given the environment and the perceived user's goal. The theory behind neural decoding is motivated by adaptive architectures whose scale and time/space complexity demand specialized computing architectures. This chapter will cover the system-level neural information processing and provide examples of the robustness of decoding between the user and the algorithms in unforeseen physical environments.

6.2 Evolution of decoders

Many groups have conducted decoding research in neuroprosthetics, and the approach has been strongly signal processing based with less emphasis on incorporating the design principles of the biologic system in the interface. The implementation path of training a decoding model has either taken an unsupervised approach by finding causal relationships in the data (Buzsáki, 2006), a supervised approach using (functional) regression (Kim et al., 2006; Sanchez and Principe, 2006, 2007), or more sophisticated methods of sequential estimation (Brown et al., 2004) to minimize the error between predicted and known behavior. These approaches, as shown in Figure 6.2, are primarily data-driven techniques that seek out correlation and structure between spatiotemporal neural activation and behavior. A mathematic or algorithmic model captures the relationships in the data, which contain variables or parameters that relate neural activity to the physiological function being decoded. Once the model is trained, the procedure is to fix the model parameters for use in a test set to validate if the physiologic function can be decoded from neural signals that were not used in the training. Some of the best known linear models that have used this architecture in the brain–machine interface (BMI) literature are the Wiener filter (FIR) (Wessberg et al., 2000; Serruya et al., 2002) and Population Vector (Helms Tillery et al., 2003), generative models (Moran and Schwartz, 1999; Taylor et al., 2002; Wu et al., 2002), nonlinear dynamic neural networks such as a time-delay neural network or recurrent neural network (Chapin et al., 1999; Sanchez et al., 2002; Gao et al., 2003), point process models (Brown et al., 2004; Eden et al., 2004; Truccolo et al., 2004), and reinforcement based models (DiGiovanna et al., 2009; Pohlmeyer et al., 2014; Bauer and Gharabaghi, 2015). A summary of each of these decoders as well as their training methodologies and practical uses is presented in Table 6.1. In general,

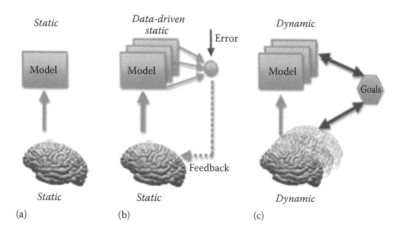

Figure 6.2 Evolution of BMI modeling. (a) First-generation BMIs with strong static assumptions for both the model and the user. (b) Data-driven approach for relaxing static modeling assumptions through the use of multiple models trained with an error response. (c) Coadaptive neuroprosthetics have a common goal for the user and model thus allowing for dynamic sharing of control.

most of the research into learning theory for training these decoders is based on adaptive signal processing and computational neuroscience. These learning rules used can be subdivided into supervised learning (where there is a "teacher"—a specific definition of a desired target and error signal) and unsupervised learning rules (where the decoder follows some set of rules, typically so that there is a classification of a set of inputs). Supervised learning algorithms include the various perceptron learning rules (delta rule, Widrow–Hoff rule, adaline rule, perceptron rule, least mean square rule, and gradient descent rule are all variations on this theme) (Haykin, 1996) and the later elaboration of these rules in the backward-propagation (back-prop) algorithm (Haykin, 1994). Unsupervised learning algorithms generally follow the viewpoint elaborated by Hebb—groups of cells that are active together wire together to form cell assemblies (Hebb, 1949). These cell assemblies can then represent objects in the world. Mathematically, the general form of these rules is the definition of sets of input weight vectors that then gradually match up with input patterns.

6.2.1 Assumptions that influence neural decoding

Figure 6.2 also illustrates the assumptions that have influenced the ongoing evolution neural decoding, where the most commonly used design in the early neuroprosthetic decoders is depicted in Figure 6.2a. Decoders of this type proved the possibility of translating thoughts into actions but are not guaranteed to work reliably for multiple tasks for multiple periods. The models that assume behavior

106

Table 6.1 Comparison of Common Neuroprosthetic Decoders

Decoder Type	Description	Training	Uses
Wiener filter	Linear combination of neural signals	Explicit error signal is needed to compute autocorrelation and cross correlation among the data	A simple and quick way to decode neural signals when a training signal is present; it is a first choice for benchmarking decoding
Population vector	Linear combination of neuronal firing rates for preferred direction of movement	Compute the preferred direction of firing for each neuron and use regression techniques to find the optimal combination of firing	A good use of neurophysiologic understanding to help seed the architecture and design of the decoder
Kalman filter	State estimate of neural signals and noise associated with them	Adaptive state estimation where the model parameters are optimized to minimize state estimation error variance	The recursive nature of this decoder enables one to weight or tune the time history of the state prediction to achieve different levels of performance
Neural network	Nonlinear combination of neural signals	Backpropagation (or one of its variants) of an error signal through the network	A good choice if it is unknown is if there is a nonlinear relationship in the neural signals used for decoding; many standard architectures are readily available
Point process model	Decoding of the fine timing of neuronal spiking activity	Two-stage process to first characterize the probability of neural spiking activity given the biological signal; the second step is to use Bayesian statistical theory to determine the most probable value of the signal given the spiking activity	A good choice to derive quantitative descriptions of how neuron populations encode information
Reinforcement based	Linear or nonlinear combination of neural signals	No error signal is required. Only a performance score (i.e., reward, no-reward) is used to reinforce or penalize model parameter updates	A good choice when there is no access to an explicit training signal; this model can continuously adapt to changes in neural or environmental signals

107

can be captured by a static input–output model and that the neural signal statistics do not change over time. Although these models can be made to work well in specific scenarios, they are not general purpose because they assume that the neural signals are stationary and that the tasks are always the same. Figure 6.2b depicts a more biologically plausible architecture with multiple models specialized for multiple functions. They can be designed to self-organize, thus improving both reliability and accuracy when trained using a supervised training or error signal. When the error response is used in training, it simplifies model building but also enforces a role of signal translator to the overall neuroprosthetic decoding system, which is incompatible with the dynamic nature of brain activity and the tasks in the environment. Here the model translates neural activity through a translating function. Using the error signal optimizes the function, but it assumes that this error signal is available for the model. This may not be the case for all functions of the brain and for all circumstances. To create a neuroprosthetic decoder capable of capturing new and complex tasks with minimal training and to accommodate the changing brain activity, a coadaptive decoder is necessary. This new design should learn online and continuously from the interaction with the user and the environment, which will replace the need for an error signal and offline learning in a training set. The only way that this can be accomplished is to create a more intelligent decoding system that enables evaluation of actions and learns from goals independently as shown in Figure 6.2c.

Using goals, a system can assign value to actions and behavior outcomes. For goal-directed behavior, it is desirable to have systems that compute with action-outcome sequences and assign high value to outcomes that yield desirable rewards. In the design of neuroprosthetic decoders, responsiveness to actions is highly desirable through the immediate update of value as soon as an outcome changes. One of the main computational goals in coadaptive decoding is to develop real-time methods for modeling and coupling the goals between the user and the decoder (to enhance symbiosis) in a variety of tasks. Here it is important to distinguish between adaptability and symbiotic coadaptation that refers to a much deeper interaction between the user and the decoder, which share control to reach common goals and learn continuously together. This approach goes beyond the simple combination of neurobiological and computation models because it elucidates the relationship between the linked, heterogeneous responses of neural systems to behavioral outcomes. A coadaptive decoder considers the interactions that influence the net benefits of behavioral, computational, and physiological strategies. Through feedback, the expression of neural intent continuously shapes the decoding model, whereas the behavior of the decoder shapes the user. A major challenge is to define appropriate computational architectures to determine the mechanistic links between the neurophysiologic levels of abstraction and behavior and to understand the evolution of neuroprosthetic usage. To determine the success of the goal-directed behavior, both the user and decoder have different measures of outcome. The prediction of success is the consequence of uncertainty in goal achievement and can be directly linked either to an inherent characteristic of the environment or to internal representations of reward in the user. The reward expectation of the user is expressed in reward centers of the brain (Schultz, 2000) and evaluates the states of environment in terms of an increase or decrease in the probability of earning reward. During the cycles of the PARC, the reward expectation of the user can be modulated by the novelty and type of environmental conditions encountered. Ideally, the goal of a coadaptive decoder is to create synergies in both outcome measures.

6.2.2 Model generalization

The primary goal in neuroprosthetic decoding is to produce the best estimates of brain function from neuronal activity that has not been used to train the model. This testing performance describes the generalization ability of the models. To achieve good generalization for a given problem, the two first considerations to be addressed are the choice of model topology and training algorithm. These choices are especially important in the design of neuroprosthetics because performance is dependent on how well the model deals with the large dimensionality of the input (many simultaneous neural signals) as well as how the model generalizes in nonstationary environments. The generalization of the model can be explained in terms of the bias-variance dilemma of machine learning (Geman et al., 1992), which is related to the number of free parameters of a model. The multiple input multiple output (MIMO) structure of neuroprosthetic decoders can have as few as several hundred to as many as several thousand free parameters. In one extreme, if the model does not contain enough parameters, there are too few degrees of freedom to fit the function to be estimated, which results in bias errors. In the other extreme, models with too many degrees of freedom tend to overfit the function to be estimated. Neuroprosthetic models tend to err on the latter because of the large dimensionality of the input.

To handle the bias-variance dilemma, one could use the traditional Akaike or BIC criteria; however, the MIMO structure of neuroprosthetic decoders excludes these approaches (Akaike, 1974). As a second option, during model training regularization, techniques that attempt to reduce the value of unimportant model parameters to zero and effectively prune the size of the model topology could be implemented (Wahba, 1990).

Neuroprosthetic decoders not only have to contend with regularization issues but also must consider ill-conditioned model solutions that result from the use of finite data sets. For example, the computation of the optimal solution for the Wiener filter involves inverting a poorly conditioned input correlation matrix that results from sparse neural firing data that are highly variable. One method of dealing with this problem is to use the pseudoinverse. However, because both conditioning and regularization are important, ridge regression (RR) (Hoerl and Kennard, 1970) can be used. For other model topologies such neural networks, weight decay (WD) regularization is an online method of RR. Both RR and WD can be viewed as the implementations of a Bayesian approach to complexity control in supervised learning using a zero-mean Gaussian prior (Neal, 1996).

A second method that can be used to maximize the generalization of a neuroprosthetic decoder is called cross validation. Developments in learning theory have shown that there is a point of maximum generalization during model training, after which model performance on unseen data will begin to deteriorate (Vapnik, 1999). After this point, the model is said to be overtrained. To circumvent this problem, a cross-validation set can be used to indicate an early stopping point in the training procedure. To implement this method, the training data are divided into a training set and a cross-validation set. Periodically during model training, the cross-validation set is used to test the performance of the model. When the error in the validation set begins to increase, the training should be stopped.

6.3 Extracting neural features as control signals

The first step in the design of any neural decoder is to select a feature of neuronal activity to use to drive the model. A good neural feature should be specific and sensitive to a cognitive condition (movement, memory, sensation, etc.). It is also desirable that the user can control (directly or indirectly) the signal with minimal training or no training at all. The choice of available brain signals and recoding methods can greatly influence the ability to extract control features, ease of clinical implementation, and operating performance. In practical terms, the designer must choose which activity to sample, what information to extract, and how to preprocess the information. The representation or spatiotemporal structure of neural activation forms a coding of intent for action in the external world. At a given moment, the neural code can be sampled as a brain state defined as the vector of values (from all recording electrodes) that describe the operating point within a space of all possible state values. The syntax or sequence of brain states must be able to support a sufficiently rich computational repertoire and must encode a range of values with sufficient accuracy and discriminability. These brain states could contain either a local or a distributed representation depending on where the signals are being collected.

It is important to note that, unlike the user's internal neural representation, the neuroprosthetic decoder representation is embodied in the computer environment. The neuroprosthetic decoder physically probes brain activity and must extract brain states creating a communication channel from the internal to external worlds. In the external world, the state representation of the decoder includes the sequence of sensory information about the environment that is relevant to goal-directed behavior.

110

6.3.1 Mistaking noise for neural signals

The level of neuroprosthetic decoding performance may be attributed to selection of electrode technology, choice of model, and methods for extracting rate, frequency, or timing neuromodulation. One of the primary goals of feature extraction for neuroprosthetics is to separate the desired neural signal from unrelated noise that may be correlated with function but is not principally the signal of interest. The types of noise encountered depend on the method of interfacing with the brain. In the case of implantable microelectrode arrays where many large populations of single neurons are acquired, a potential noise source could be the fraction of neurons that are firing in the background but do not directly modulate for the desired functional neuroprosthetic use. For example, consider a motor neuroprosthetic where 100 neurons are recorded from the primary motor cortex. Out of that population, one-third may change their firing properties at any given moment to control movement of the arm. However, there are the remaining two-thirds of the neurons that are still active but do not modulate for that specific task. These neurons contribute additional variance to the decoder model that must be either suppressed or handled in a way that does not contribute to the variance of the decoder control.

Another example of these aspects of noise is encountered in the use of the electroencephalogram (EEG) methods for interfacing with the brain. When using EEG, artifacts from eye movements can be introduced into the neural signals. They may be correlated with particular behaviors (i.e., looking at an object one is seeking to reach). However, they are not directly neural in nature and would be a surrogate for pure neuroprosthetic control. Similarly, muscular activity from the sides of the head as well as the facial muscles can contribute to changes in EEG activity. Again, these are often mistaken for neural activity because they are often correlated with a user's behavior. Because EEG is a macroscopic approach to neural interfacing, it also aggregates neuronal activity for a large volume of tissue. This lack of specificity leads to the neuronal signals on each channel, including a mixture of neural activity from the neurons of interest as well as background EEG. Because these signals are mixed together on each single channel, they find it much more difficult to contend with a required advanced signal processing compared to the single neuron microelectrode neural interfaces. In the latter, placing different neural activities on each separate channel allows the designer to just turn off that channel to minimize the influence of noise. This is not directly possible with other types of interfaces that include mesoscopic or macroscopic activity.

6.3.2 Example of neural features

Three domains of signal features are commonly used as inputs to neural decoding models. They include processing of modulations of neuronal activity in the frequency, temporal, and spatial domains. Examples of local field potential (LFP)

111

spectrograms, spike–LFP coherence, principal component analysis (PCA), and event-related potentials (ERPs) are shown next.

6.3.2.1 LFP spectrograms Frequency domain analysis is commonly achieved via Fourier transform, wavelet transform, or autoregressive modeling. Figure 6.3a shows a simplified raw neural recording signal. This signal is comprised of multiple brain oscillatory activities. For illustration purposes, consider that this signal is just the sum of two different frequency sine waves. Although this is simplified, the design of neural features for neuroprosthetics often involves seeking out specific oscillations that are related to a specific behavioral function. One tool for identifying these oscillations is to perform a Fourier analysis (Oppenheim et al., 1999). Here the Fourier analysis reveals two peaks in the spectrum, each corresponding to the frequencies contained in the elementary signals making up the raw neural signal. If this were real neural recordings, finding peaks that only appear during specific behaviors would reveal an indicator that a viable neural signal has been identified. By contrast, be aware that

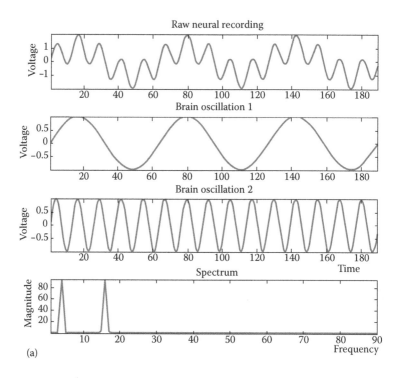

Figure 6.3 (a) Conceptual decomposition of spectral content in neural recordings.
(Continued)

112

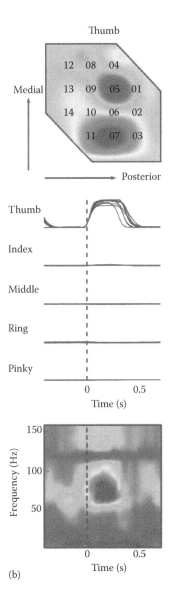

(b)

Figure 6.3 (Continued) (b) Spectrogram of ECoG neural activity time synchronized with movement. (Adapted from Wang, W., A. D. Degenhart, J. L. Collinger, R. Vinjamuri, G. P. Sudre, P. D. Adelson, D. L. Holder et al., Human motor cortical activity recorded with micro-ECoG electrodes, during individual finger movements. *Conference Proceedings: Annual International Conference of the IEEE Engineering in Medicine and Biology Society IEEE Engineering in Medicine and Biology Society Conference*, 586–9, 2009. doi: 10.1109/IEMBS.2009.5333704.)

peaks in the spectrum can also appear at frequencies associated with line noise (i.e., 60 Hz) and its harmonics. It is up to the designer to verify the validity of the spectral or noise information. One method of performing this validation is shown in Figure 6.3b, where the displacement (i.e., movement) of fingers is time synchronized with the spectrogram of neural recordings derived from the electrocorticogram (ECoG) of a human subject. The spectrogram or short-time Fourier transform is similar to the spectrum in Figure 6.3a; however, it is computed in many sliding windows over time. Each slide of time in Figure 6.3b produces a spectrum indicating which frequencies of brain oscillations are active. In this example, there is an increase in the activity between 50 and 100 Hz only when the subject's thumb moves and is not active at any other time. This indicates these frequencies may be viable for neuroprosthetic control for the thumb. Once features like these have been identified, time domain techniques such as band-pass filtering can be applied to the raw neural recordings to isolate specific frequencies as inputs to neural decoders. In this example, a band-pass filter in the 50- to 100-Hz range would produce a time-varying neural control feature. Lastly, it is also important to consider the spatial components of where these signals appear and how noise influences them. Spatial filters such as common average referencing (CAR) can be used to remove noise that appears across many channels (Ludwig et al., 2009). The CAR is not a computational complex method, but it can have drastic improvements on performance. To apply it, simply average all the recordings on every electrode site of interest and use it as a reference for each recording channel.

6.3.2.2 Spike rates and spike–LFP coherence The mechanism of neural connectionism and communication involves neuronal signaling across a synapse (i.e., between its dendritic input and the axonal output), which is mediated through a series of changes in cell membrane potential. Molecular and voltage-gated signaling occurs at the synapses that are chemical–electrical connectors between the axon of presynaptic neurons and one of many dendritic inputs of a given neuron. The axon of a firing presynaptic neuron transfers the transient current and subsequent action potential (spike) through a series of changes in membrane potential to its many synaptic endings. At the synapse, ions are released to the attached dendrite, locally increasing the electrical potential at the dendrite's distal portion. This voltage is slowly integrated with contributions from other dendritic branches and propagated until the cell body. For individual neurons, when the electrical potential at the base of the dendritic tree crosses a threshold that is controlled by the cell body, the neuron fires its own action potential that is transferred through its axonal output to other postsynaptic neurons. Through this signaling procedure, there are two types of electric fields produced by neurons: waves of dendritic synaptic current that establish weak potential fields as the dendritic synaptic current flows across the fixed extracellular cortical resistance, and trains of action potentials on the axons that can be treated as point

114

processes. The generation of these "spike events" is shown by the vertical lines in Figure 6.4, and the aggregate of extracellular dendritic activity gives rise to the LFP. A third representation of neuronal signals includes the firing rate, which is simply the sum of action potentials in a unit of time. In Figure 6.4, each example window of time is 1 s; therefore, the firing rates are 5 and 4 Hz, respectively. The spike–LFP coherence is a frequency domain representation of the similarity of dynamics between a spike train and the LFP voltage fluctuations. Figure 6.4a shows a low spike–LFP coherence because the temporal occurrence of the spike does not align with the peaks of the particular frequency of the LFP wave. Figure 6.4b shows high spike–LFP coherence. Note that the single measurement of the firing rate only provides one point of reference and thus is not the best for computing the coherence in this example.

6.3.2.3 Principal component analysis The volume of data (in space and time) of multiple neural signals for use as inputs to the decoder study can make signal analysis overwhelming in the design of neuroprosthetics. Ultimately selecting neural signals with the largest variance that is correlated with behavior produces the best decoders. Changes in the variance of neural activity signal important events for decoding. The use of PCA on neural recording can be used to linearly project the data in a way that reduces the number of parameters while finding the directions of maximal variance. The approach consists of finding a projection of the covariance of the neural data such that the covariance of the projected data is equal to the eigenvalues of the original neural recordings. To process any neural signal via PCA, first compute an eigendecomposition on the covariance of the neural recordings. Next, use the eigenvectors to filter the data. This process transfers a set of correlated variables into a new set of uncorrelated variables that are in a space of lower dimensionality. Because the new axes are orthogonal and represent the directions with maximum variability, PCA can be viewed as a rotation of the original axes to new positions in the space defined by original variables. The

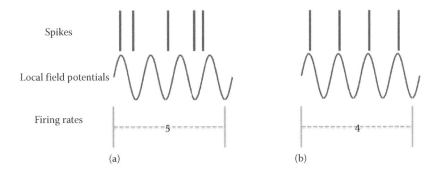

Figure 6.4 The relationship between spikes, LFP, and neuronal firing rates for a unit of time (1 s). (a) Low spike–LFP coherence. (b) High spike–LFP coherence.

first PC retains the greatest amount of variation in the sample, whereas the nth PC retains the nth greatest fraction of the variation in the data. PCA can be used in a wide variety of neuroprosthetic applications, including spike sorting and finding variance in neuronal firing rate and field potential recordings.

6.3.2.4 Event-related potentials For mesoscopic and macroscopic signals such as ECoG and EEG, ERPs can be used as a signature of cognition. They signal a massive communication among multiple brain regions. In engineering terms, they can be thought of as a kind of "brain impulse response" to an internal stimulus. As shown in Figure 6.5, the shape of the ERP is well known, spatially distinct, and can be easily reproduced among individuals. The ERP is characterized by negative and positive peaks that occur at different times with respect to a stimulus or event. It is important to note that the negative and positive terms are swapped with respect to the y-axis because of the polarity of measured brain signals. In this example, the P300 occurs as a negative peak approximately 300 ms poststimulus. The P300 signals a rare task-relevant event over Cz in the 10- to 20-electrode arrangement in EEG. It is used very commonly in BMI experiments and is elicited via "oddball" stimuli (He et al., 2001; Brunner et al., 2011). The N100-P200 complex is a preattentive response occurring 100 to 200 ms over sensory areas. Although these signals are reproducible, the problem is that it is normally much smaller than the ongoing EEG or ECoG activity (i.e., the signal-to-noise ratio [SNR] is negative). To contend with the negative SNR, signal averaging is commonly used. Using the assumption that the ERP appears as a transient in white Gaussian noise, aligning the response and averaging across trials can increase the SNR by $\dfrac{1}{\sqrt{N}}$, where N is the number of trials. This is approach is commonly used in neuroprosthetic design but has three shortcomings: it is not real time, it

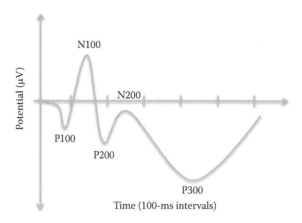

Figure 6.5 Major components of event-related potentials.

116

assumes that the shape of the ERP is the same, and it assumes that the latency is constant. These shortcomings should be carefully considered in the context of the specific neuroprosthetic application to determine if they are suitable.

6.4 Examples of neuroprosthetic decoders

Two examples are given here to illustrate how to build a direct link between the brain and a tool that interacts with the environment. The first example is the most fundamental neural decoder and is the Wiener filter. The Wiener filter will be developed in the context of a trajectory brain–machine interface as would be encountered when learning how to control a robotic arm to follow a movement trajectory. The second example is an advanced decoder, which is developed in the context of a coadaptive brain–machine interface. The objective here is to illustrate how to design new computational architectures that can serve as assistants to facilitate control and can serve as an equalizer to share the burden of learning or generating the appropriate neuroprosthetic commands.

6.4.1 Fundamental example—Wiener decoder

The most fundamental neural decoder assumes that there exists a linear mapping between the desired behavioral function and the neuronal activity. In this model, the delayed versions of the neural activity, $\mathbf{x}(t - l)$, are the bases that construct the output signal. Figure 6.6 shows the topology of the MIMO Wiener filter, where the output y_j is a weighted linear combination of the l most recent values of neuronal input $\mathbf{x}$ given by $y_j(t) = \mathbf{W}_j\mathbf{x}(t)$ (Haykin, 1996). Here y_j is the decoded signal, and it can be defined to be any behavioral function such as movement (x, y, z Cartesian positional information of a limb). The model parameters are updated using the optimal linear least squares solution that matches the Wiener solution. The Wiener solution is given by $\mathbf{W}_j = \mathbf{R}^{-1}\mathbf{P}_j = E(\mathbf{x}^T\mathbf{x})^{-1}\, E(\mathbf{x}^T\mathbf{d}_j)$, where $\mathbf{R}$ and $\mathbf{P}_j$

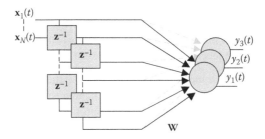

Figure 6.6 Linear Wiener filter topology. Each neuronal input $\mathbf{x}_N$ contains a tap-delay line with l taps. The output $\mathbf{y}$ is a weighted ($\mathbf{W}$) sum of the neuronal inputs.

are the autocorrelation and cross-correlation functions, respectively, and $\mathbf{d}_j$ is a known behavioral function (i.e., x, y, z Cartesian positional information of a limb). This signal is derived from direct observations of the outcomes of behavior by a neuroprosthetic user.

The autocorrelation matrix $\mathbf{R}$ and the cross-correlation matrix $\mathbf{P}$ can be estimated directly from the data using either the autocorrelation or the covariance methods (Haykin, 1996). In this MIMO problem, the autocorrelation matrix is not Toeplitz, even when the autocorrelation method is employed. One of the real dangers of computing the Wiener filter solution for neuroprosthetic decoders is that $\mathbf{R}$ may not be full rank (Sanchez et al., 2004). Instead of using the Moore–Penrose inverse, a regularized solution substituting $\mathbf{R}^{-1}$ by $(\mathbf{R} + \lambda I)^{-1}$, where λ is the regularization constant estimated from a cross-validation set, may be more appropriate. Effectively, this solution corresponds to RR (Hoerl and Kennard, 1970). The computational complexity of the Wiener filter is high for the number of input channels used in our experiments. For 100 neural channels using 10 tap delays and three outputs, the total number of weights is 3,000. This means that one must invert a $1,000 \times 1,000$ matrix every N samples, where N is the size of the training data block.

An alternative to the Wiener solution is to use a search procedure to find the optimal solution of $\mathbf{W}$ using gradient descent or Newton-type search algorithms. The most widely used algorithm in this setting is the least mean squares (LMS) algorithm, which uses stochastic gradient descent (Haykin, 1996). For real data, the normalized LMS algorithm is a good choice and given by $\mathbf{W}_j(t+1) = \mathbf{W}_j(t) + \dfrac{\eta}{\|x(t)\|}\mathbf{e}_j(t)\mathbf{x}(t)$, where η is the step size or learning rate, $e(t) = d_x(t) - y_x(t)$ is the error, and $\|x(t)\|$ is the power of the input signal contained in the taps of the filter. Using the normalized LMS algorithm, the model parameters are updated incrementally at every new sample, and thus the computation is greatly reduced. The step size must be experimentally determined from the data. One may think that the issues of nonstationarity are largely resolved because the filter is always being updated, tracking the changing statistics. However, during testing, the desired response is not available, so the weights of the filter have to be frozen after training. Therefore, the normalized LMS (NLMS) is still subject to the same problems as the Wiener solution, although it may provide slightly better results when properly trained (when the data are not stationary, the LMS and the Wiener solution do not necessarily coincide; Haykin, 1996).

Linear filters trained with mean square error provide the best linear estimate of the mapping between neural firing patterns and hand position. Although the solution is guaranteed to converge to the global optimum, the model assumes the relationship between neural activity and behavior, which may not be the case. Furthermore, for large input spaces, including memory in the input introduces many extra degrees of freedom to the model, hindering generalization capabilities.

6.4.1.1 Decoding hand trajectory The decoding performance of the Wiener filter is demonstrated on a 3-D hand reaching task. Here a primate performed a right-handed reach to food and subsequently placed the food in its mouth. Neuronal firings, binned (added) in nonoverlapping windows of 100 ms, are directly used as inputs to the Wiener decoder. The primate's hand position is used as the desired signal. The training data consisted of 20,010 consecutive time bins (2001 seconds) of recordings.

In testing, all model parameters were fixed, and 3000 consecutive bins (300 s) of novel neuronal data were fed into the models to predict new hand trajectories. Figure 6.7 shows the output in the test set with 3-D hand position for three reaching movements. From the plots, it can be seen that qualitatively the Wiener decoder does a fair job at capturing the movements and nonmovements in the trajectory. However, the Wiener decoder cannot maintain the peak values of the hand position.

By far, the easiest model to implement is the Wiener filter. With its quick computation time and straightforward linear algebra, it is clearly an attractive decoder choice for neuroprosthetics. Its function can also be explained in terms of simple weighted sums of delayed versions of the ensemble neuronal firing (i.e., it is correlating neuronal activity with hand position). However, from the trajectories shown, the output is noisy and does not accurately capture the details of the movement. These errors may be attributed to the solution obtained from inverting a poorly conditioned autocorrelation matrix and also to the number of free parameters in the model topology.

6.4.2 Advanced example—reinforcement learning decoder

Research has revealed multiple factors that influence neural decoding accuracy on even short timescales (hours to days). For example, performance can be enhanced or degraded by the quantity, type, and stability of the neural signals acquired (Wessberg et al., 2000; Carmena et al., 2003; Mehring et al., 2003, 2004; Paninski et al., 2004; Sanchez et al., 2004; Lebedev et al., 2005; Santhanam

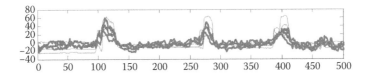

Figure 6.7 Testing performance of a Wiener decoder for a 3-D reaching task. Here the y-axis is displacement and the x-axis is time in 100-ms bins. Bold lines are the x, y, and z trajectories estimated by the decoder. Fine traces are the true x, y, and z trajectories of the subject.

119

et al., 2006), the effects of learning and plasticity (Taylor et al., 2002; Carmena et al., 2003; Lebedev et al., 2005; Ganguly and Carmena, 2009), the availability of physical signals for training the neural decoders (Taylor et al., 2002; Hochberg et al., 2006; Simeral et al., 2011), and the duration of decoder use (Wessberg et al., 2000; Pohlmeyer et al., 2007; Ganguly and Carmena, 2009). These conditions create a dynamic substrate from which neuroprosthetic designers and users need to produce stable and robust decoding performance if the systems are to be used for activities of daily living and increase independence for the BMI users.

Two of the particularly significant challenges to neural decoders include how to create a decoder when a user is unable to produce a measureable physical output to map to the neural activity for training the decoder, and how to maintain performance for both long and short timescales when neural perturbations (inevitably) occur. For neuroprosthetics that use chronically implanted microelectrode arrays in the brain, these perturbations include the loss or addition of neurons to the electrode recordings, failure of the electrodes themselves, and changes in neuron behavior that affect the statistics of the decoder input firing patterns over time.

In this example, neural decoders based on reinforcement learning (RL) (Pohlmeyer et al., 2012; Mahmoudi et al., 2013) are illustrated. RL is an interactive learning method designed to allow systems to obtain reward by learning to interact with the environment, and which has adaptation built into the algorithm itself using an evaluative scalar feedback signal (Sutton and Barto, 1998). As with supervised adaptation methods, these decoders can adapt their parameters to respond to user performance. Unlike supervised adaptation methods, they use a decoding framework that does not rely on known (or inferred) targets or outputs (such as kinematics) as a desired response for training or updating the decoder. Therefore, they can be used even when such information is unavailable (as would be the case in highly unstructured BMI environments), or when the output of the current decoder is random (e.g., an uncalibrated BMI system, or when a large change has occurred within the decoder input space), because they use a scalar qualitative feedback as a reinforcement signal to adapt the decoder. In learning theory, reward learning is very different from supervised learning because the decision maker is not told explicitly how to adapt, rather it receives feedback about what actions lead to the most reward. Several studies have shown that RL can be used to control basic neuroprosthetic systems using EEG signals (Iturrate et al., 2010; Matsuzaki et al., 2011) and neuron activity in rats (DiGiovanna et al., 2009; Mahmoudi and Sanchez, 2011). The RL neural decoder shown here is based on the theory of associative RL (Mahmoudi et al., 2013).

The RL neural decoder is tested under three basic conditions: an absence of explicit kinetic or kinematic training signals, large changes (i.e., perturbations) in the neural input space, and control across long periods. Two marmoset monkeys use the reinforcement learning brain–machine interface (RLBMI) to control a robot arm during a two-target reaching task. Two robot actions are used to emphasize the relationship between each specific robot action, feedback signal, perturbations, and resulting RLBMI adaptation. The RLBMI parameters are initially seeded using random numbers, with the system only requiring a simple "good/bad" training signal to quickly provide accurate control that can be extended throughout sessions spanning multiple days. Furthermore, the RLBMI automatically adapts and maintains performance despite very large perturbations to the BMI input space. These perturbations include either sudden large-scale losses or additions of neurons among the neural recordings.

6.4.2.1 Subjects and neural data acquisition In this example, two (PR and DU) marmoset monkeys (*Callithrix jacchus*) control a robot arm during a two-choice reaching task. Each monkey is implanted with a 16-channel tungsten microelectrode array (Tucker Davis Technologies, Alachua FL) in the motor cortex, targeting arm and hand areas. Neural data are acquired using a Tucker Davis Technologies RZ2 system (Tucker Davis Technologies, Alachua, FL). Each array is re-referenced in real time using a CAR composed of that particular array's 16 electrodes (if an electrode failed, it was removed from the CAR) to improve SNR (Ludwig et al., 2009). Neural data are sampled at 24.414 kHz and band-pass filtered (300 Hz–5 kHz). Action potential waveforms are discriminated in real time based on manually defined waveform amplitudes and shapes. The recorded neural data include both multineuron signals and well-isolated single neuron signals (collectively referred to here as neural signals), which are used equivalently in all real-time and offline tests. Neural signal firing rates are normalized (between −1 and 1) in real time by updating an estimate of the neural signals' maximum firing rates during each experimental trial.

6.4.2.2 Actor–critic RLBMI control architecture The actor–critic RLBMI architecture demonstrated here is described in detail by Mahmoudi et al. (2013). Briefly, actor–critic systems are characterized by the actor and critic modules, as shown in Figure 6.8a. The actor interacts with the environment by selecting system actions given a specific input state (here neural states). The critic provides an evaluative feedback regarding how successful the actions were in terms of some measure of performance, and which is used to refine the actor's state to action mapping. The actor is a fully connected three-layer feedforward neural network (Figure 6.8b) that used a Hebbian update structure (Mahmoudi et al., 2013). The actor input (X) is a vector (length n) of the spike counts for each of the n motor cortex neural signals during a 2-s window after the go cue of each trial. The network is designed to contain only five hidden nodes and two output nodes (one for each of the two robot reaching movements). The output of each hidden node (OutH$_i$) is a probability of firing (−1 to 1) computed using a hyperbolic tangent function, and in which $\mathbf{WH}_i$ is the synaptic weight vector between node i

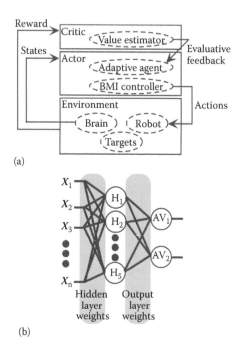

(a)

(b)

Figure 6.8 Brain–machine interface control architecture using actor–critic reinforcement learning. (a) In this architecture, the actor interacts with the environment by selecting actions given input states. The critic is responsible for producing reward feedback that reflects the actions' impact on the environment and which is used by the actor to improve its input to action mapping capability (here the adaptive agent). (b) The actor used here is a fully connected three layer feedforward neural network with five hidden (H_j) and two output (AV_j) nodes. The actor input ($\boldsymbol{X}$) was the normalized firing rates of each motor cortex neural signal. Each node was a processing element which calculated spiking probabilities using a tanh function, with the node emitting spikes for positive values.

and the n inputs ($\mathbf{b}$ is a bias term); $\text{OutH}_j = \tanh([\mathbf{X} \quad \mathbf{b}] * \mathbf{WH}_j)$. The output nodes determine the action values (AV) for each of the j possible robot actions; $AV_j = \tanh([S(\mathbf{OutH}) \quad \mathbf{b}] * \mathbf{WO}_j)$. $S(\mathbf{OutH})$ is a sign function applied to the hidden layer outputs (positive values become +1, negative values become −1), and $\mathbf{WO}_j$ is the weights matrix between output j and the hidden layer. The robot action with the highest action value is implemented each trial. The actor weights are initialized using random numbers and are updated using the critic feedback (f):

$$\Delta\mathbf{WH} = \mu_H * f * ([\mathbf{X} \quad \mathbf{b}]^T * (S(\mathbf{OutH}) - \mathbf{OutH}))$$

$$+ \mu_H * (1-f) * ([\mathbf{X} \quad \mathbf{b}]^T * (1 - S(\mathbf{OutH}) - \mathbf{OutH}))$$

122

and

$$\Delta \mathbf{WO} = \mu_O * f * ([\mathbf{OutH} \quad \mathbf{b}]^T * (S(\mathbf{AV}) - \mathbf{AV}))$$
$$+ \mu_O * (1-f) * ([\mathbf{OutH} \quad \mathbf{b}]^T * (1 - S(\mathbf{AV}) - \mathbf{AV}))$$

Feedback is +1 if the previous action selection is rewarded and −1 otherwise. This update equation is composed of two terms that provide a balance between the effects of reward and the punishment on the network parameters. Under rewarding conditions, the first term contributes to the changes in the synaptic weights, whereas in the case of punishment, both terms will affect the weight update. After convergence to an effective control policy, the output of the node tends to the sign function, and thus the adaptation will stop automatically (Mahmoudi et al., 2013). In this example, an "ideal" critic is used, which always provided accurate feedback. However, such perfect feedback is not intrinsically assumed by this RL architecture, and there are many potential sources of the feedback that include deriving critic signals directly from the brain. $S()$ is again the sign function, and μ_H and μ_O are learning rates of the hidden (0.01) and output (0.05) layers, respectively. The update equations are structured so that the local input–output correlations in each node are reinforced using a global evaluative feedback, hence Hebbian RL. The decoder here is applied to a two-state problem, but this architecture and update equations can be directly applied to multistep and multitarget problems, even while still using a binary feedback signal (Mahmoudi et al., 2013).

6.4.2.3 Neural-controlled robot reaching task The BMI task requires the monkeys to make reaching movements with a robot arm to two different spatial locations to receive food rewards (Figure 6.9a). The monkeys initiate trials by placing their hand on a touch sensor for a randomized hold period (700–1200 ms). The hold period is followed by an audio go cue, which coincides with the robot arm moving to the start position. Simultaneous with the robot movement, an LED spatial target on either the monkeys' left ("A" trials) or right ("B" trials) is illuminated. Before the real-time BMI experiments, the monkeys are trained to manually control the robot movements by either making or withholding arm movements. During those training sessions, the monkeys move the robot to the A target by reaching and touching a second sensor and to the B target by keeping their hand motionless on the touchpad, and they were rewarded for moving the robot to the illuminated target. The differences in the neuron firing rates shown by the rasters in Figure 6.9b illustrate how this had trained the monkeys to associate changes in motor activity with moving the robot to the A target, and static motor activity to B target robot movements. In the real-time test, the robot movements are determined directly from the monkeys' motor cortex activity using the actor–critic RL algorithm previously described. A and B trials are presented in a pseudorandom order of roughly equivalent proportions. The monkeys are

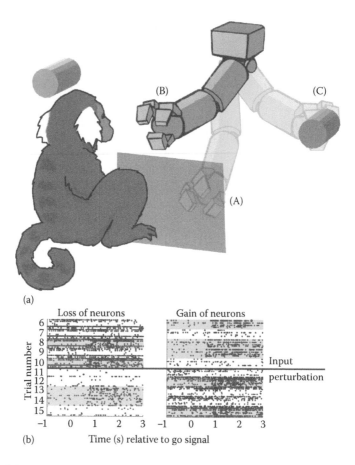

Figure 6.9 Two target robot reaching task. (a) A robot arm moved out from behind an opaque screen (position A) and presented its gripper to the monkey (position B). A target LED on either monkey's left (A trials) or right (B trials) is illuminated to indicate the goal reach location. The monkeys' motor cortex activity is used to either move the robot to the A or B target. (b) Examples of the spike rasters for all the neural signals used as inputs to the RLBMI during experiments, which tested the effects of neural signals being lost or gained. Data are shown for trials 6–10 (which preceded the input perturbation) and trials 11–15 (which followed the input perturbation).

immediately given food rewards at the end of trials only if they move the robot to the illuminated LED target.

6.4.2.4 RLBMI stability when initialized from random initial conditions During real-time closed-loop robot control experiments, the parameter weights of the RLBMI are initialized with random values, with the RLBMI learning effective action mappings through experience. Performance is quantified as the percentage of trials in which the target is achieved. In addition to these closed-loop real-time experiments, a large number of offline "open-loop" Monte Carlo simulations are run to exhaustively confirm that the RLBMI is robust in terms of its initial conditions, i.e., that convergence of the actor weights to an effective control state during the real-time experiments had not been dependent on any specific subset of initialization values.

6.4.2.5 RLBMI stability during input space perturbations: loss or gain of neuron recordings For BMI systems to show truly stable performance, nonstationarities or other changes in the input space should not adversely affect performance. Although some changes of the input space can be beneficial, such as neurons changing their firing pattern to better suit the BMI controller (Taylor et al., 2002; Carmena et al., 2003; Lebedev et al., 2005; Jarosiewicz et al., 2008; Ganguly and Carmena, 2009; Ganguly et al., 2011; Chase et al., 2012), large changes in the firing patterns of the inputs that dramatically remove the input space from that which the BMI had been constructed around are significant problems for BMIs. Such perturbations can result from neurons appearing or disappearing from the electrode recordings, a common occurrence in electrophysiology recordings.

In several closed-loop BMI sessions, the inputs to the decoder are deliberately altered to test the RLBMI's ability to cope with large-scale input perturbations. These perturbations are imposed after the initial learning period so that the RLBMI had already adapted and gained accurate control of the robot before input perturbation.

6.4.2.6 RLBMI decoder stability over long periods This example illustrates how the RLBMI performs when it is applied in closed-loop mode across long periods. These contiguous multisession tests consist of a series of robot reaching experiments for each monkey. During the first session, the RLBMI is initialized using a random set of initial conditions. During the follow-up sessions, the RLBMI is initialized from weights that it had learned from the prior session and then continued to adapt over time.

The impact of input perturbations is also tested during the contiguous multisession experiments. In these experiments, a random half of the motor neural signals is selected (the same signals in each test), and in those perturbation experiments, the firing rates of the selected inputs are set to zero.

6.4.2.7 Actor–critic RLBMI control of robot arm The actor–critic RLBMI effectively controlled the robot reaching movements. Figure 6.10 shows a typical closed-loop RLBMI experimental session (PR). Figure 6.10a shows that the algorithm converged to an effective control state in less than five trials, after which the robot consistently made successful movements. The algorithm was initialized using small random numbers (±0.075) for the parameter weights. Figure 6.10b

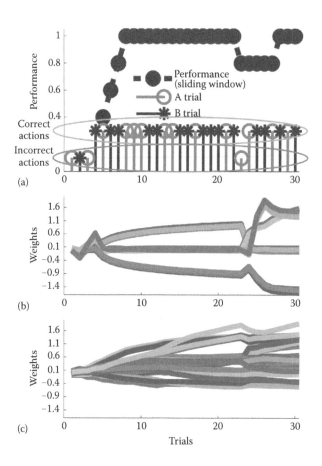

Figure 6.10 The RLBMI accurately learns to control the robot during closed-loop BMI experiments. (a) Stems indicate the sequence of the different trials types (O = A trials, * = B trials) with the stem height, indicating whether the robot moves to the correct target (taller stem) or not (shorter stem). Panels b and c show how throughout every trial the RLBMI system gradually adapts each of the individual weights that connected the hidden layer to the outputs (b) as well as all the weights of the connections between the inputs and the hidden layer (c), as the RLBMI learns to control the robot.

shows the gradual adaptation of the weight values of the two output nodes as the algorithm learns to map neural states to robot actions (Figure 6.10c shows a similar adaptation progression for the hidden layer weights). The weights initially change rapidly as the system moves away from random explorations, followed by smooth adaptation and stabilization when critic feedback consistently indicates good performance. Larger adaptations occur when the feedback indicated an error has been made.

126

The RLBMI system is very stable over different closed-loop sessions, robustly finding an effective control policy regardless of the parameter weights' initial conditions. Figure 6.11 shows that during the closed-loop robot control experiments, the RLBMI controller selects the correct target in approximately 90% of the trials (light gray bars mean ± standard deviation; DU: 93%, five sessions; PR: 89%, four sessions, significantly above chance [0.5] for both monkeys, $p < 0.001$, one-sided t-test). Similarly, Figure 6.11 dark gray bars shows that the open-loop initial condition Monte Carlo simulations yield similar accuracy as the closed-loop experiments, confirming that the system converges to an effective control state from a wide range of initial conditions (DU: 1000 simulations; PR: 700, significantly above chance [0.5] for both monkeys; $p < 0.001$, one-sided t-test).

A surrogate data test is used to confirm that the RLBMI decoder was using the monkeys' brain activity to control the robot arm and not some other aspect of the experimental design. These tests involve additional open-loop simulations in which the order of the different trial types recorded during the real-time experiments is preserved while the order of the recorded motor cortex neural data is randomly reshuffled, thus destroying any consistent neural representations associated with the desired robot movements. Despite the decoder's adaptation capabilities, Figure 6.11 (black bars) shows that the RLBMI system was not able to perform above chance levels under these conditions (DU: 1000 simulations; PR: 700, $p \sim 1$, one-sided t-test), demonstrating that the RLBMI is unable to accomplish the task without the direct connection between the motor cortex command signals and the desired robot actions that had been recorded during the real-time experiments.

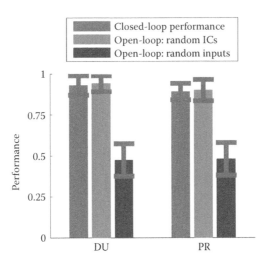

Figure 6.11 The RLBMI decoder accuracy for robot control.

127

6.4.2.8 RLBMI performance during input space perturbations The RLBMI quickly adapts to compensate for large perturbations to the neural input space. Figure 6.12 gives the accuracy (mean and standard deviation) of the RLBMI decoder within a five-trial sliding window across the trial sequence of both closed-loop BMI experiments (DU: light gray dashed line and error bars, four sessions) and open-loop simulations (DU: gray line and panel, 1000 simulations; PR: dark gray and panel, 700 simulations). Figure 6.12a shows the RLBMI performance when a random 50% of the inputs were lost after trial 10 (vertical black bar). By trial 10, the RLBMI has already achieved stable control of the robot, and it has readapted to the perturbation within five trials, restoring effective control of the robot to the monkey. The inset panel in Figure 6.12a contrasts the mean results of the RLBMI simulations (solid lines) against simulations in which a static neural decoder (dashed lines, specifically a Wiener decoder) is used to generate the robot action commands. The Wiener decoder initially performs quite well, but the input perturbation causes a permanent loss of performance. Figure 6.12b shows that the RLBMI system effectively incorporates newly "found" neural signals into its input space. This input perturbation again occurs after the 10th trial (vertical black bar). Prior to that point, a random 50% of the RLBMI inputs had had their firing rate information set to zero. In both the closed-loop BMI experiments and

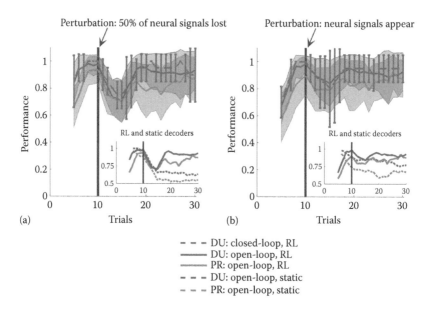

Figure 6.12 The RLBMI quickly adapts to perturbations to the neural input space. These perturbations included both the loss of 50% of the neural inputs (a) as well as when the number of neural signals detected by the neural recording system doubled (b).

128

open-loop simulations, the system again has adapted to the input perturbation within five trials. By comparison, a static decoder (Wiener) is not only unable to take advantage of the newly available neural information but in fact shows a performance drop after the input perturbation (Figure 6.12b inset panel; RLBMI: solid lines; static Wiener: dashed lines).

Both the losses of 50% of the recorded inputs and the abrupt appearance of new information among half the recordings represent significant shifts to the RLBMI input space. Figure 6.13a contrasts the change in the available information between the neural signals with losses of varying quantities of neural signals (boxes; DU: solid; PR: hollow). By the time 50% of the inputs have been lost, more than half of the information had been lost as well. Abrupt input shifts of this magnitude would be extremely difficult for any static neural decoder to overcome. It is thus not unexpected that the static Wiener decoder (circles; DU: solid; PR: hollow) nears

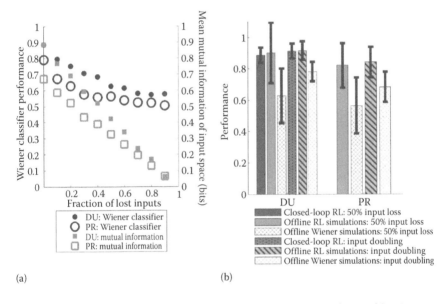

(a) (b)

Figure 6.13 Input perturbations cause significant performance drops without adaptation. (a) The effect of different fractions of neural signals being lost on the performance of a nonadaptive neural decoder (Wiener decoder), relative to the average information available from the neural inputs (DU: 1000 simulations; PR: 700 simulations). Panel (b) shows how the RLBMI adapted to large input perturbations (50% loss of neural signals and doubling of neural signals) during both closed-loop experiments (signals lost: dark gray; new signals appear: dotted gray; 4 experiments) and the offline simulations (signals lost: light gray; new signals appear: diagonal gray; DU: 1000 simulations; PR: 700 simulations), resulting in higher performance than the nonadaptive Wiener decoder (hatched boxes, one-sided t-test, $p \ll 0.001$).

129

chance performance by this point. Any decoder that does not adapt to the change would show similar performance drops. Figure 6.13b contrasts the average performance (trials 11–30) of the RLBMI after perturbations during both closed-loop experiments (neural signals lost: dark gray; new neural signals appearing: dotted gray) and open-loop simulations (neural signals lost: light gray; new neural signals appearing: diagonal gray) against the performance of the static Wiener decoder (hatched bars; neural signals lost: dark; new neural signals appearing: light). The RLBMI performance is significantly higher than the nonadaptive Wiener decoder (one-sided t-test, $p \ll 0.001$; DU: 1000 simulations; PR: 700 simulations).

6.4.2.9 Potential benefits of using RL algorithms for BMI decoders Adaptive and interactive algorithms based on RL offer several significant advantages as BMI controllers over supervised learning decoders. First, they do not require an explicit set of training data to be initialized; instead, they are computationally optimized through experience. Second, RL algorithms do not assume stationarity between neural inputs and behavioral outputs, making them less sensitive to failures of recording electrodes, neurons changing their firing patterns due to learning or plasticity, neurons appearing or disappearing from recordings, or other input space perturbations. These attributes are important considerations if neuroprosthetics are to be used by humans over long periods for activities of daily living.

6.4.2.10 The RL decoder does not require explicit training data The RLBMI decoding architecture does not require an explicit set of training data to create the robot controller. BMIs that use supervised learning methods require neural data that can be related to specific BMI behavioral outputs (i.e., the training data) to calibrate the BMI. In many BMI experiments that have used healthy nonhuman primate subjects, the training data outputs were specific arm movements that accomplished the same task for which the BMI would later be used (Wessberg et al., 2000; Serruya et al., 2002; Taylor et al., 2002; Carmena et al., 2003; Santhanam et al., 2006; Li et al., 2009, 2011; Pohlmeyer et al., 2009; Sussillo et al., 2012). Although those methods were effective, gaining access to this type of training data is problematic when considering paralyzed BMI users. Other studies have found that carefully structured paradigms that involve a BMI user first observing (or mentally imagining) desired BMI outputs, followed by a process of refinements that gradually turn full BMI control of the system over to the user, can provide training data without physical movements (Velliste et al., 2008; Hochberg et al., 2012; Collinger et al., 2013). Although these methods have been effective, they require carefully structured BMI paradigms so that assumed BMI outputs can be used for the calibration. The RLBMI decoder shown here avoids such issues. Calculating the parameters of the RLBMI architecture never requires relating neural states to known (or inferred) system outputs; however, the system starts controlling the robot with random parameters, which are then gradually adapted given feedback of current performance. Thus, the robot made random movements when the system was initialized, but the RLBMI is able to quickly (typically within two to four trials) adapt the parameters to give accurate control (~90%) of the robot arm. This adaption only requires a simple binary feedback. Importantly, the same RLBMI architecture used here can be directly applied to tasks that involve more than two action decisions, while still using the exact same weight update equations. This means that the system can be readily extended to more sophisticated BMI tasks while still only requiring the same type of binary training feedback (Mahmoudi et al., 2013).

6.4.2.11 The RL decoder remains stable despite perturbations to the neural input It is important that changes in a BMI's neural input space do not diminish the user's control, especially when considering longer periods where such shifts are inevitable (Carmena et al., 2005; Santhanam et al., 2007; Dickey et al., 2009; Chestek et al., 2011; Fraser and Schwartz, 2012). For example, losses and gains of neurons are very common with electrophysiology recordings using chronically implanted microelectrode arrays: electrodes fail entirely, small relative motions between the brain and the electrodes cause neurons to appear and disappear, and even the longest-lasting recording arrays show gradual losses of neurons over time either from the tissue encapsulation of the electrodes or from the gradual degradation of the electrode material (Turner et al., 1999; Prasad et al., 2012). Although some changes in neural input behavior can be beneficial, such as neurons gradually adopting new firing patterns to provide a BMI user greater control of the system (Taylor et al., 2002; Carmena et al., 2003; Lebedev et al., 2005; Jarosiewicz et al., 2008; Ganguly and Carmena, 2009; Ganguly et al., 2011; Chase et al., 2012), large and/or sudden changes in neuron firing patterns will almost always reduce a BMI user's control if the system cannot compensate, as is with the static Wiener decoder.

Although input losses may be an obvious adverse perturbation to BMI systems, the appearance of new neurons is also a significant input perturbation: when the representations of new neurons overlap with neurons that were already being used as BMI inputs, this causes the previous inputs to appear to have acquired new firing patterns, thus perturbing the BMI's input space. Such appearances could be a particular issue in BMI systems that rely on action potential threshold crossings on a per electrode basis to detect input activity (Fraser et al., 2009; Gilja et al., 2012; Sussillo et al., 2012). Finally, BMIs that cannot take advantage of new sources of information lose the opportunity to compensate for losses of other neurons.

Currently, most BMI experiments avoid the issue of large changes in input neurons on BMI performance because the experimenters reinitialize the systems on a daily basis at least (Hochberg et al., 2006, 2012; Velliste et al., 2008; Simeral et al., 2011; Gilja et al., 2012; Orsborn et al., 2012; Collinger et al., 2013). However, it is important for practical BMI systems to have a straightforward method of dealing with neural input space perturbations that are not a burden on the BMI user and do not require such daily recalibrations. The RLBMI controller shown here does not require the intervention of an external technician (such as an engineer or caregiver) to recalibrate the BMI after changes in the input space. Rather, it automatically compensates for input losses in a way that the RLBMI adapted and suffered only a transient drop in performance despite neural signals disappearing from the input space. In situations in which the algorithm had set the silent channel parameter weights very close to zero, or in which the activity of new channels was relatively low, the addition of the new neural signals would have had little impact on performance until the RLBMI controller reweighted the perturbed inputs appropriately to effectively use them. Conversely, during the input loss tests, there would be a higher probability that dropped inputs had had significant weight parameters previously attached to their activity, resulting in a more obvious impact on overall performance when those neural signals were lost.

The RLBMI architecture is intended to balance the adaptive nature of the decoder with the brain's learning processes. Understanding the intricacies of these dynamics is important for the study of brain function and BMI development. Numerous studies have shown that neurons can adapt to better control BMIs (Taylor et al., 2002; Carmena et al., 2003; Lebedev et al., 2005; Jarosiewicz et al., 2008; Ganguly and Carmena, 2009; Ganguly et al., 2011; Chase et al., 2012). RL adaptation is designed so that it does not confound these natural learning processes. RL adaption occurs primarily when natural neuron adaptation is insufficient, such as during the initialization of the BMI system or in response to large input space perturbations. The mutual optimal adaptation of both the brain and the neural decoder can offer highly effective and robust BMI controllers.

6.4.2.12 Obtaining and using feedback for RLBMI adaptation   The ability of the RLBMI system to appropriately adapt itself depends on the system receiving useful feedback regarding its current performance. Thus, how accurate the critic feedback is and how often it is available directly impact the RLBMI's performance. The experimental setup shown here assumed an ideal case in which completely accurate feedback was available immediately after each robot action. Although such a situation is unlikely in everyday life, it is not essential for RL that feedback always be available and/or correct, and there are many potential methods by which feedback information can be obtained.

The RLBMI architecture presented here does not intrinsically assume perpetually available feedback but rather only needs feedback when necessary and/or convenient. If no feedback information is available, then the update equations are simply not implemented, and the current system parameters remain unchanged. Because feedback information does not depend on any particular preprogrammed training paradigm, but rather simply involves the user contributing good/bad information during whatever task for which they are currently using the BMI, this makes the system straightforward to update by the user whenever is convenient, and they feel the RLBMI performance has degraded. Finally, other RL algorithms are designed specifically to take advantage of only infrequently available feedback by relating it to multiple earlier actions that were taken by the system and which ultimately lead to the feedback (Pilarski et al., 2011).

There are a wide variety of potential options for the RLBMI user to provide critic feedback to the system, including using reward or error information encoded in the brain itself. While assuming that ideal feedback is available after each action may not be practical for real BMI systems, the fact that the necessary training feedback is just a binary "good/bad" signal (even when the system is expanded to include more than two output actions) that only needs to be provided when the user feels the BMI performance needs to be updated leaves many options for how even a user suffering from extreme paralysis could learn to provide critic feedback. For example, the user could use a breath puff system, vocal cues, or any sort of small residual movement or EMG signal that can be reliably evoked. Furthermore, error-related signals characteristic to EEG, ECoG, or other recording methods could be employed as well (Llera et al., 2011; Hoffmann and Falkenstein, 2012; Kreilinger et al., 2012; Milekovic et al., 2012; Spuler et al., 2012; Geng et al., 2013; Prins et al., 2013). An exciting option that would place the smallest burden on the BMI user would be to automatically decode feedback information regarding the BMI's performance directly from the brain itself, perhaps from learning or reward centers such as the nucleus accumbens, anterior cingulate cortex, prefrontal cortex, etc. (Schultz, 2000). Giving the BMI user access to a straightforward method of providing feedback will enable them to use the BMI system effectively for long periods without outside interventions by engineers or caregivers despite inevitable changes to the inputs. This will greatly increase the practicality of BMI systems by increasing user independence.

Exercises

1. Draw a neural decoding model topology for a neuroprosthetic decoder. Label the inputs, outputs, training signals, time delays, and weights as would be used for a neuroprosthetic system that restores 3-D movement of an upper extremity.
2. What are several of the problems and/or factors that affect performance when using input/output engineering models for neuromotor decoding applications?
3. You are interested in training a neural decoder for a brain function in which a desired signal is not available (i.e., no explicit error signal to adapt the decoder weights). Describe in detail the alternative options for how the decoder could be trained and implemented.

Chapter 7 Principles of stimulation

Neuroscientists are novices at deception. Magicians have done controlled testing in human perception for thousands of years.

Teller

Learning objectives

- Understand how electrical signals can artificially be injected into the nervous system to induce the modulation of neuron activity.
- Master the variables involved in delivering electrical stimulation to tailor neuroprosthetic interfaces for any application.
- Determine which methods of stimulation will be safe or unsafe for any neuroprosthetic system design.

7.1 Introduction

Nobel Prize winner Gerald Edelman once stated that "The brain is embodied and the body is embedded" (Edelman, 2004). The coupled and persistent interactions between the brain, body, and environment ultimately give rise the conscious perception of one's self. The resultant activity and development of the brain from the body or environment are all driven by stimuli. In nature, these stimuli could be sounds that are transduced by your ears, images transduced by your eyes, or touch transduced by your skin. The design of neuroprosthetic systems that have the capability of "writing in" information to the nervous system fundamentally change the concept of transducing signals from the environment and into the brain. The transduction of stimuli is no longer through natural sensory organs but through engineered systems. At its core, the concept of a stimulus for a neuroprosthetic is any perturbation in an organism's internal or external environment that results in changes in the nervous system physiology. On a macro scale, the physiologic responses to natural stimuli or neuroprosthetic electrical current

135

could be the creation of a muscle contraction, the activation of pyramidal neurons in the primary sensory cortex to produce the sensation of touch, or the creation of an electric field at the cell and molecular level of abstraction to repair or replace damaged tissue. As electricity moves through conductive media of the body, responses can be evoked in cells and tissues. The effects can be direct and occur along the path of current flow in the local vicinity of the electrode. Alternatively, the effects can result distally from the site of stimulation through a systems-based response of the neural tissue propagating the stimuli.

7.2 Nerve responses to electrical current

The interaction between electric fields and excitable neural tissue gives rise to the responses needed to write information into the brain in neuroprosthetic applications (Figure 7.1). The electric fields generated by currents applied to the extracellular space of neural tissue can be calculated by solving Maxwell equations (Plonsey, 1969). Although there are detailed formulations for this biophysical relationship, the overarching factors among the basic neural physiology and electrical stimulation are presented here. The mechanism of action for neuronal responses is caused by how electrodes interfaced with the nervous system generate electric fields that influence a cell membrane's voltage-sensitive permeability. The site of current injection in Figure 7.1a (block 2) creates the largest upswing in the membrane potential compared with other sites along the neuron. Figure 7.1b shows the relative differences in membrane potential if they were recorded as a function of the distance away from the site of stimulation. Neurons, in their natural state, produce unequal distribution of charged ions on each side of the membrane, which creates a potential difference between the interior and the exterior of a cell. This creates a resting potential of -65 mV as shown in Figure 7.1c. The neuron tries to maintain this electrochemical gradient as its normal homeostatic environment. To do this, the neuron contains active transport pumps that continually move Na^+ from inside the cell to the outside, and balances this positive charge movement by moving K^+ to the inside. This active transport produces an electrical gradient with positively charged ions outside and negatively charged ions inside the cell membrane. Axonal excitation by applied current, as shown in Figure 7.1d, through an electrode is effected when the transmembrane current generated by the electrode depolarizes the membrane sufficiently to activate the sodium channels located in high densities at various locations in the cell. To create the transmission of an impulse in a nerve, the resting membrane potential must be reduced below the threshold level. Once activated, the sodium current will further increase until the membrane reaches an unstable fixed point. This is the point at which a full action potential will develop. Once started, the action will propagate unattenuated either smoothly along the unmyelinated fibers or discretely at the nodes of Ranvier of the myelinated fibers.

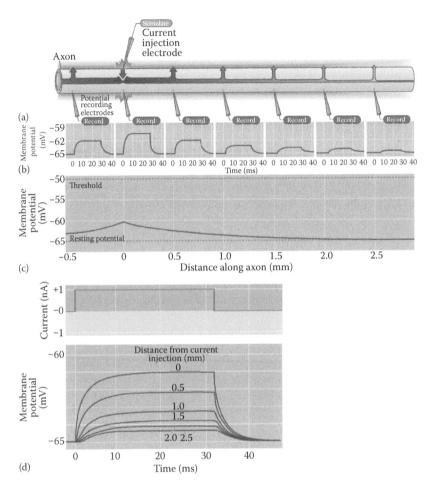

Figure 7.1 The effects of injected current on membrane potential. Sites local to the stimulation produce the largest changes in voltage. Membrane voltage falls off exponentially as a function of distance along the axon of the neuron. (a) Experimental setup. (b) Membrane potential responses at the sites corresponding to the microelectrodes. (c) Relationship between the membrane potential and distance along axon. (d) Superimposed responses (from b). (Adapted from Purves, D., G. J. Augustine, D. Fitzpatrick, L. C. Katz, A.-S. LaMantia, J. O. McNamara, and S. M. Williams. *Neuroscience.* Sunderland, MA: Sinauer Associates, 2001.)

7.3 Strength–duration curves

Precise delivery of stimulation to neural tissue is one of the hallmarks of restoring function in neuroprosthetic applications. The specificity of delivering neural stimuli depends on the amplitude and duration of the pulses of electrical activity delivered. Strength–duration curves are used to represent the threshold for depolarization of a nerve fiber or aggregate neural network. As shown in Figure 7.2, the exponential shape of the curve is the most critical aspect of the relationship because it shows how the threshold current for producing action potentials decreases with increasing stimulus pulse width. At long pulse widths, the current needed to excite the tissue is minimum. This minimum current is called the rheobase, whereas the chronaxie describes the length of time required for a current of twice the intensity of the rheobase current to produce tissue excitation. Higher currents are needed with short pulses to produce an equivalent stimulus compared with long duration pulses. In general, strength–duration curves can be defined by the following formulation: $I_{th} = I_{RH} (1 + (T_c/PW))$, where I_{th} is the threshold current, PW is the pulse width, I_{RH} is the rheobase current I_{RH}, and T_c is the chronaxie time.

7.3.1 Activation order

The threshold for stimulation can be selective depending on the neural structure being activated. For example, in the peripheral nervous system, nerves depolarize in the same order starting with the sensory nerves, motor nerves, pain nerves, and muscle fibers. This is based on the cross-sectional diameter (i.e., large-diameter nerves depolarize first) and the location of the nerve with respect to the stimulating electrode (i.e., superficial nerves depolarize first for surface stimulation). In the design of neuroprosthetics, one must pay special attention to the relationship between the recruitment order obtained by engineering and the activation order in biology. This is especially true for the neuroprosthetic control of peripheral muscles. Naturally, the

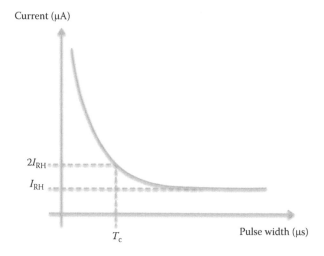

Figure 7.2 Strength–duration curves provide the relationship between the applied current pulse and duration of the stimulus pulse to produce an action potential.

138

body will activate small fibers for small forces. If larger forces are needed, the larger fibers will then be activated. However, as discussed previously, large fibers have a lower threshold of activation compared with small fibers, and when using electrical stimulation in a neuroprosthetic, the recruitment order is reversed. This can create a problem for natural neuroprosthetic control if strategies are not used to obtain the right activation order (Gorman and Mortimer, 1983; Grill and Mortimer, 1995).

7.3.2 Adaptation

Adaptations to repeated stimulation can affect the strength and duration needed to activate neural structures. The frequency and duration of the stimuli can affect how the nervous system and the individual learns to process neuroprosthetic stimuli. As a result, more intense stimulation is needed to achieve the same effect. This can have a major effect on the power requirements for device design. Two forms of adaptation are commonly observed in neuroprosthetic applications: habituation and accommodation. In the case of habituation, repeated stimulation can induce a situation in which the nervous systems begin to reduce or cease response to nonmeaningful repetitive stimulation. Accommodation refers to increases in the depolarization threshold of a neuron after repeated stimulation. The mechanisms for accommodation are complex and involve the dynamics of activation and inactivation of ion channels (Hodgkin and Huxley, 1952). Both habituation and accommodation can be prevented by varying the stimulation applied to the nervous system. This can come in the form of the specific pattern of activation as well as the duration that it is applied.

7.4 Current flow

A detailed understanding of the flow of current to/from the stimulating electrode and through biological tissues enables precise targeting and activation of neural circuits. In the setup of a neural stimulator system, electrons flow from the device anode (negative terminal) through the neural tissue and to the device cathode (positive terminal). In contrast, ion flow occurs only within the tissues. Here, negative ions in solution flow toward the anode and away from the cathode. Positive ions in solution flow toward the cathode and away from the anode. There are multiple factors that affect the flow of ions or electrons in biological tissue. They include the use of alternating versus direct current, tissue impedance, current density, and electrode placement. In the next sections, each of these concepts will be explored with respect to their effect on the activation of neural tissue.

7.4.1 Current density

Current density refers to the amount of current per unit area of biological tissue. In general, the higher the current density, the more intense the effect of stimulation

139

becomes. To compute the current density for any electrode, calculate the current setting on the stimulator device and divide that number by the area of the uninsulated portion of the electrode that is in contact with the biological tissue. There are multiple electrode configurations that affect the current density, as shown in Figure 7.3 (Freeman, 1975). Here, each block represents an electrode and the lines represent the current that is in the tissue. Figure 7.3a is a "bipolar technique" where one electrode is connected to the anode and the other to the cathode of the stimulator. Here, there are equal current densities and equal effects under each electrode.

7.4.2 Electrode size

Electrode size plays a major role in current density because it appears in the denominator of the current density calculation. To illustrate this effect, Figure 7.3c shows a configuration in which one electrode is larger than the other. Here the larger spacing of the lines beneath the larger electrode shows the smaller current density due to the distribution of current paths. This configuration can also be considered a "monopolar technique" in which the effects of stimulation are concentrated under the smaller electrode. It is considered to be the "active" electrode compared with the larger contact.

7.4.3 Tissue impedance

Although the images in Figure 7.3 show a uniform substrate through which the current flows, tissue impedance can create inconsistencies in the resulting current

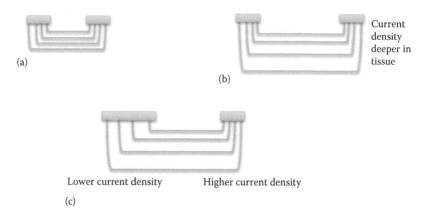

Figure 7.3 Resulting changes in current density. Blocks represent electrodes and lines represent current. (a) A reference pair of electrodes. (b) Increased spacing between electrodes results in current density deeper in the tissue. (c) Increased side of the electrode creates a lower current density.

density. Tissue impedance refers to the resistance of tissue to the passage of electrical current. Relatively, bone and fat are considered to be high-impedance whereas nerve, blood vessels, and muscle are low-impedance. Current flow will find the path of least resistance. This concept is important in neuroprosthetic design for real biological tissue because current in these scenarios do not flow through the shortest path from electrode to electrode because the path of least resistance is not necessarily the shortest path. Variations in blood vessels, bone, and nerve cytoarchitecture can create a multitude of paths for current to flow. Designers of neural-stimulating prosthetics should consider the arrangement of low and high impedance tissues to determine the most effective way to activate the neural tissue.

7.4.4 Electrode arrangements

Figure 7.3b shows how the proximity of the anode and cathode electrodes can affect current density. The relative location of the electrode determines the number of parallel paths of current. The farther apart the electrodes are, the more parallel paths are formed. This has the effect of increasing the current density in deeper tissues. However, there is a trade-off as more current is required to produce effects as the number of paths increases.

7.5 Current types

The application of electrical stimulation to the nervous system involves a combinatorial explosion of options for the currents themselves. For neuroprosthetic applications, the goal is always to achieve the most precise stimulation for the desired functional effect. As described in Section 7.3, the precise activation of neurons is related to the interaction between strength and duration. With just these two aspects, the designer of a neuroprosthetic can construct trains of pulses that have amplitude and pulse timing properties that can be tailored to mimic the natural activation of neural tissue or create new artificial activation that can produce a variety of beneficial and harmful effects on the electrodes themselves and responses in neural tissue. A primary concern in the design is how the application of stimulating current affects the oxidation and reduction chemical reactions that govern the neural electrode interface. There are two primary mechanisms of charge transfer at the electrode–electrolyte interface and they include faradaic and nonfaradaic reactions (McIntyre, 2011; Merrill, 2011). Nonfaradaic reactions occur when the amount of charge is small. These reactions do not involve electron transfer and only include redistribution of ions in the electrochemical solution. Faradaic reactions involve larger charges and facilitate oxidation and reduction reactions by transferring electrons at the electrode–electrolyte interface. Alternating versus direct current plays a big role in these processes. The biggest effect is the ability of constant direct current to cause, over time, irreversible chemical changes at the electrode–tissue interface. Depending on the choice of electrode material (Pr, Ir, IrO, W, etc.) and the solution of ions it is in contact with, the application of current can produce oxidation/reduction reactions that have numerous products. These include the production of hydrogen gas bubbles at the electrode tip that can damage neural tissue, the deposition of metal ions into solution that can be toxic, corrosion of the metal surface, and changes in pH due to the reduction of water (Johnson et al., 2005; Merrill, 2011). Oxidation and reduction reactions are dynamic processes, which can be affected by the polarity of the pulses applied. If the level of stimulation produces a reversible reaction, changing the polarity can cause the reactants of the chemical process to be reformed from the products. Reversibility depends heavily on the amount of electron transfer, the kinetics of charged particles and the mass transport of reactants at the electrode–tissue interface.

7.5.1 Stimulation-induced tissue damage

The fundamental design of neuroprosthetic stimulation systems must keep the electrode action within a potential window where irreversible faradaic reactions do not produce damaging effects to the electrode or physiological system. Figure 7.4 aggregates results from multiple studies to define safe and unsafe stimulation parameters in the space spanned by charge and charge density (Merrill et al., 2005). Depending on the stimulation parameters and the size of the electrode, designers must consider how microelectrodes and macroelectrodes balance the

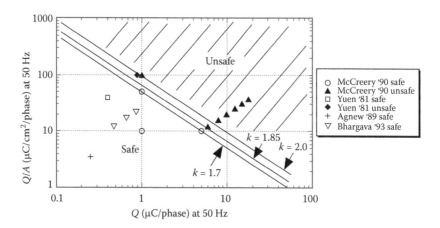

Figure 7.4 Relationship between charge (Q) and charge density (Q/A) for safe stimulation. (Adapted from Merrill, D. R., M. Bikson, and J. G. R. Jefferys. Electrical stimulation of excitable tissue: Design of efficacious and safe protocols. *J. Neurosci. Methods* 141(2):171–98, 2005.)

142

total charge per pulse and charge density to safely stimulate. Selecting parameters that operate in the unsafe zone will produce tissue damage through the creation of toxic electrochemical reaction products at a rate greater than that which can be tolerated by the physiological system. In addition to the parameters described in Figure 7.4, a second consideration includes tissue damage caused by the mass action theory of intrinsic biological tissue that is overstimulated. The damage occurs from the change in local environment because of the induced hyperactivity of many neurons firing or neurons firing for an extended period (Freeman, 1975).

7.5.2 Pulsed currents

Pulsed currents can be used to control what types of reactions occur at the electrode–tissue interface. Monophasic currents as shown in Figure 7.5 only contain one direction of flow followed by periods of noncurrent flow. Biphasic currents contain bidirectional flow of electrons that occur on both the positive and negative sides of the baseline. These can be used to facilitate the reversible reactions described previously. Figure 7.5 also illustrates the components of monophasic and biphasic stimulation configurations. Here, the pulse duration is the time from when the pulse rises from the baseline to the point where it terminates on the baseline. The amount of time that the current is flowing relative to the time it is not flowing is called the duty cycle (DC). The DC = "ON"/("ON + OFF") × 100. For example, if the stimulus is on for 2 seconds and is off for 5 seconds, the DC = 2/(2 + 5) × 100 = 28.6%. Stimulation can include multiple phases or components of the pulse that appear on one side of the baseline. Monophasic currents only have one phase and it is equivalent to the pulse duration. More complex pulses such as biphasic stimulation have two phases. As described earlier, the duration of these phases can influence the activation sequence of neural tissue. In addition to the basic phases,

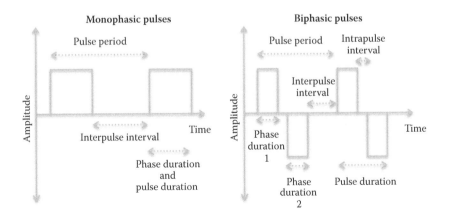

Figure 7.5 Summary of common electrical stimulation parameters.

143

interpulse interval describes the time between the end of one pulse and the start of the next pulse. Changing the interpulse interval has the effect of modulating the pulse frequency. Intrapulse intervals are brief interruptions of current flow. In general, these are shorter than the interpulse interval. Changing the intrapulse interval can influence the total charge delivered by the pulse. Combining the phase duration, intrapulse interval, and interpulse interval produces the pulse period, which is the total amount of time from the start of one pulse to the start of the next pulse. Stimulation of neural tissue is commonly reported in terms of frequency (Hz). The pulse frequency is defined as the number of times the stimulation occurs per second. In addition to the basic properties of individual stimulation pulses, neuroprosthetic designers also report stimulation in terms of pulse trains or bursts. In this configuration, a series of individual pulses will be delivered followed by a period of no stimulation. After some time, the train of stimulation will repeat again. The period between stimulations is called the interburst or intertrain interval.

7.5.3 Effects of pulsed current types

Pulsed currents can be symmetrical, as shown in Figure 7.5, where the positive and negative components are mirror images of each other. Thus, there are no net positive or negative charges that could affect the electrode or electrolyte. Variations of phase duration 1 with respect to phase duration 2 can produce either balanced or unbalanced asymmetrical pulsed current types. For balanced asymmetrical pulses, the shape allows for both positive and negative effects but the net positive and negative sum of the charge remains zero. Unbalanced asymmetrical pulses also have positive and negative effects but the imbalance in the positive and negative charges result in a net sum of charge that is nonzero over time. This type of stimulation can produce oxidation and reduction tissue effects.

7.6 Example applications

The fundamental concepts of stimulation for the development of neuroprosthetic devices have been developed here. The factors discussed are general in nature and, to be effectively applied, need to be tailored to the specific application, neural target site, and electrode interface chosen. The following list illustrates the wide range of applications and the variations encountered when making design choices.

- Stimulation for the restoration of the sense of touch (Tabot et al., 2014): Microelectrode array (Utah) implanted chronically into the primary sensory cortex and stimulation is applied directly to the neurons in the layered cortex. Stimulation is persistent only for the duration while the finger is in contact with an object and the pattern of stimulation varies by the type of sensation (tactile, roughness, slip, etc.).

144

- Stimulation of the spinal cord to restore movement to the lower limbs (Angeli et al., 2014): ECoG style electrodes are placed epidurally on the spinal cord. Stimulation pulses must traverse the dura and into the neural circuits of the spinal cord. The stimulation pattern is active during walking and is repetitive for the stereotypical phases of gait.
- Stimulation of muscles to restore the ability to reach and grasp (Peckham and Knutson, 2005): Epimysial or intramuscular electrodes are used to activate muscles through functional electrical stimulation (FES). Stimulation is commonly triggered for binary responses (contracting vs. rest). For noninvasive FES, stimulation pulses must traverse the skin, fat, vasculature, muscle, and nerve. Activating the appropriate recruitment order is critical for fine control of force.
- Stimulation of the hippocampus to restore the formation and recall of memory after traumatic brain injury (Hampson et al., 2013): Microelectrodes implanted into the CA1, CA2, and CA3 of the hippocampus. Stimulation parameters must be tuned to the specific spatiotemporal codes involved in the time-varying sequence of memory function.
- Stimulation of deep brain structures for alleviating the symptoms of Parkinson's disease (Lozano and Mahant, 2004): Mesoscopic electrodes are implanted into deep brain structures such as the subthalamic nucleus. Stimulation parameters must be tuned to each subject to achieve an effect. Once derived, the stimulus settings are typically fixed for a certain duration on and off. Frequent clinical office visits are necessary to finely adjust the parameters to account for accommodation and habituation effects.

Exercises

1. Compare and contrast the advantages and disadvantages of current mode versus voltage mode stimulators.
2. How might biphasic pulses be modified to achieve other effects of stimulation, or to eliminate adverse effects?
3. Compute the duty cycle for a neural stimulator that is on for 40 seconds and off for 20 seconds.
4. Describe the order in which nerves and muscle depolarize in response to stimulation. Be sure to include sensory, motor, pain, and muscle fibers.
5. For brain stimulation applications, there is a nonlinear relationship between the pulse duration and the threshold current that are necessary to stimulate a neural element. Draw a strength–duration curve, label the axes, chronaxie, and rheobase currents. Describe, in a few sentences, the relationship between the rheobase, chronaxie, and how specific values lead to the activation of myelinated axons, cell bodies, and dendrites.

Chapter 8 Application
Brain-actuated functional electrical stimulation for rehabilitation

Jan Scheuermann: I can move it up. And straight down. And left and right, and diagonally. I can close it. And open it. And I can go forward and back.
Scott Pelley: That is just the most astounding thing I've ever seen.
Scott Pelley: Can we shake hands?
Jan Scheuermann: Sure.
Jan Scheuermann: And I can do a fist bump if you'd like.
Scott Pelley: That's amazing.

60 Minutes
Breakthrough: Robotic Limbs Moved by the Mind (2012)

Learning objectives

- Understand user priorities for people living with paralysis and how neuro-prosthetics can restore function.
- Specify rehabilitation strategies and testing metrics for motor neuroprosthetics.
- Design the components of a closed-loop, brain-actuated functional electrical stimulation system for restoring movement to the upper extremities.

8.1 Introduction

In the United States, approximately 270,000 people are living with a spinal cord injury (SCI), with approximately 12,000 new cases occurring each year. The primary causes are motor vehicle accidents (39%), falls (28%), acts of violence (15%), and sports (8%) (National SCI Statistical Center, February 2012). Trauma to the spinal cord causes a disruption in the pathways connecting the brain to the muscles, resulting in muscle weakness and paralysis. Other common symptoms include loss of sensation, muscle spasticity, sexual dysfunction, breathing problems, autonomic dysreflexia, and loss of bladder and bowel function.

The level and extent of injury determines which muscles are affected and the amount of voluntary control that remains. Damage to the cervical spinal cord often results in paralysis of both the upper and the lower limbs, and is termed either tetraplegia or quadriplegia. In contrast, damage to the thoracic, lumbar, or sacral cord results in paralysis of the lower limbs only, which is called paraplegia. The spinal cord segments affected by the injury can be determined by systematically testing the dermatomes and myotomes on both sides of the body. The American Spinal Injury Association (ASIA) publishes the International Standards for Neurological Classification of Spinal Cord Injury, which is revised every few years, most recently in 2011 (Kirshblum et al., 2013). The ASIA examination tests each dermatome and myotome separately to determine sensory and motor levels, single neurological level of injury, completeness of injury, impairment scale grading (A to E), sensory scores, motor scores, and the zone of partial preservation, as shown in Figure 8.1.

Complete cervical level SCIs, which account for 40.8% of new cases since 2005, result in total paralysis of hand muscles (Kirshblum et al., 2013). The loss of voluntary control over the hands results in an inability to perform many activities of

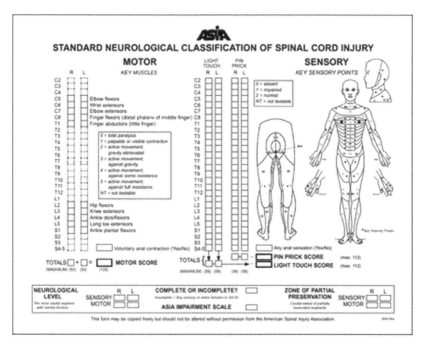

Figure 8.1 ASIA standard neurological classification of spinal cord injury.

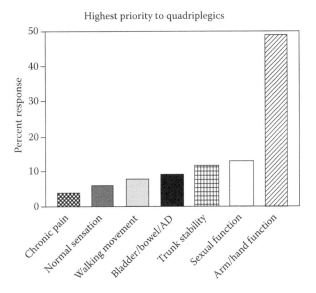

Figure 8.2 Function that would most improve the quality of life for people with tetraplegia. (Adapted from Anderson, K.D. Targeting recovery: Priorities of the spinal cord-injured population. *J Neurotrauma* 21(10):1371–83, 2004; Consideration of user priorities when developing neural prosthetics. *J Neural Eng* 6(5):055003. doi: 10.1088/1741-2560/6/5/055003, 2009.)

daily living, which in turn leads to dependence on caregivers and family. When asked to identify the function that would most dramatically improve their life, 48.7% of people with tetraplegia ranked arm/hand function as their highest priority, as shown in Figure 8.2 (Anderson, 2004, 2009). Restoration of hand function would offer those with high-level SCIs an increased level of independence and greatly improved quality of life.

8.2 Hand rehabilitation strategies

After spinal cord injury, a variety of adaptive changes can occur in the nervous system and they can affect central or peripheral targets. The brain is a dynamic organ and the lack of inputs or outputs to it can result in denervation of peripheral muscles and related brain reorganization. Studies of brain function have revealed that cortical areas, which were responsible for movement and sensation of body parts before the injury, can be reorganized or taken over by the function of areas not affected by the injury (Wall and Egger, 1971; Levy et al., 1990; Bruehlmeier et al., 1998; Raineteau and Schwab, 2001). Changes can be measured in terms of the location and intensity of activation patterns within the motor cortex, suggesting

149

a postinjury reorganization of motor control (Green et al., 1998; Kokotilo et al., 2009). Neuroplasticity is believed to play a major role in this process. Although neuroplasticity can lead to maladaptive changes, this phenomena can also be exploited during rehabilitation after SCI.

Many approaches have been used to restore hand and arm function after injury and they include rehabilitation (Hoffman and Field-Fote, 2007, 2010; Beekhuizen and Field-Fote, 2008), brain–computer interfaces (BCIs) (Pfurtscheller et al., 2000, 2003), and neurorobotics (Page et al., 2011). In all of the approaches to restore hand and arm function, the goal is to better engage voluntary control and counteract maladaptive brain reorganization that results from nonuse (Hoffman and Field-Fote, 2006). For these reasons, there is a need for new therapies that will help restore motor abilities and rehabilitate the motor cortex. Standard rehabilitation augmented with new developments from the study of neuroprosthetics could provide a new combined therapy approach (Daly and Wolpaw, 2008; Daly et al., 2009) for motor cortex rehabilitation and to alleviate motor impairments. The appeal of combined approaches to rehabilitation is that they can engage top-down control from the central nervous system (CNS) and couple it with bottom-up peripheral therapy. Such an approach would utilize the normal motor circuits and prevent the cortex from forming maladaptive connections because of the lack of outputs that connect the CNS to the peripheral nervous system (PNS) because of the SCI. To counteract this, a neuroprosthetic can bypass the injury and reconnect the brain to peripheral muscles via functional stimulation. Reconnecting the brain to the body will alleviate motor impairments and provide exercise to the muscles. Combining BCIs with the activation of paralyzed muscle by electrical stimulation is one example of how a combined therapy could provide a unique approach to motor cortex rehabilitation by enabling the user to actively control extremities during rehabilitation with brain activity (Daly and Wolpaw, 2008).

8.3 Fundamentals of functional electrical stimulation

Functional electrical stimulation (FES) is the application of electrical stimulation to weak or paralyzed muscles to produce an action (Peckham and Knutson, 2005). For more than 50 years, FES has been used extensively for the restoration of bladder and bowel function, respiratory and cardiac pacing, therapeutic exercise, and reanimation of skeletal muscle. After incomplete SCI, long-term application of FES has been shown to facilitate voluntary grasping function (Popovic et al., 2011).

After traumatic SCI, motor neuron damage and death is common and can extend multiple levels above and below the initial site of injury. Previous studies have reported complete denervation in 39% of triceps muscles and 15% of thenar muscles in subjects with complete cervical SCI (Thomas and Zijdewind, 2006). Muscles become denervated when they lose their nerve supply and become difficult to activate by electrical stimulation. People with extensive or complete denervation of the target muscles are generally excluded from FES applications, as the current threshold for initiating an action potential in nerve is much lower than in muscle, as shown in Figure 8.3.

FES can be delivered to muscle in a variety of ways. Surface systems are the least invasive, as the stimulating electrodes are placed over the motor point of a muscle. Surface systems are relatively inexpensive and easily reversible, but consistent electrode placement and activation of deep muscles can be difficult. In addition, pain receptors in the skin can be activated, making stimulation painful if sensory pathways remain intact. Fully implantable, intramuscular electrodes are most commonly used in FES systems intended for long-term use. All the components of the system are implanted inside the body. The stimulator is typically implanted in the chest or abdomen and connected to the intramuscular electrodes through subcutaneous cabling. Power and control signals are transmitted by a radiofrequency telemetry link from an external control unit. Although both surface and implantable systems have been used with success in people with SCI.

For hand opening, the extensor muscles that contribute to the movement include the extensor digitorum profundus, extensor digitorum superficialis, extensor pollicis longus, and extensor pollicis brevis. To close the hand, the contributing flexor muscles include the flexor digitorum superficialis, flexor digitorum profundus, and thenar muscles (Kilgore and Peckham, 1993; Mangold et al., 2005). Muscles can be activated either individually by placing small electrodes over the motor points of each muscle or more grossly by placing large electrodes across groups of synergistic muscles. Contraction force can be modulated by varying the total charge delivered to the muscle.

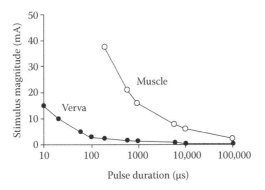

Figure 8.3 Strength duration curves for nerve and muscle in cat tibialis anterior muscle. (From Thomas Mortimer, J. Motor Prostheses. Handbook of Physiology, The Nervous System, Motor Control. *Comprehensive Physiology* Supplement 2: 155–87, 2011. doi: 10.1002/cphy.cp010205.)

151

In the application of surface stimulation to the muscles of the hand, pulse durations of 200 to 400 µs and frequencies of 20 to 40 Hz are commonly used (Doucet et al., 2012). Frequency is most commonly modulated at the beginning and end of each contraction, with ramp times of 1 to 3 seconds (Doucet et al., 2012). To produce full extension of the fingers (hand opening) in able-bodied subjects, typical pulse amplitudes range from 15 to 20 mA (Chiou et al., 2009). To achieve full flexion of the fingers (hand closing) in uninjured subjects, pulse amplitudes of 20 to 26 mA are typically required (Chiou et al., 2009). Higher pulse amplitudes are required to produce hand movements in paralyzed muscles, which are weaker and more fatigable than healthy muscles. In previous studies with SCI subjects, currents necessary for hand opening and closing vary, but are generally less than 40 mA (Chiou et al., 2009). To determine the optimal stimulation parameters for each subject, particularly those with chronic SCI, a systematic evaluation of the current hand function, excitability, and strength of the target muscle groups is necessary.

A variety of different control signals have been used to activate FES systems. Logical command signals can be used to simply turn the stimulation on or off. Proportional command signals offer graded control of stimulation, so that contractions of different strengths can be achieved. To be acceptable to an end user, the command signal task must be easy to perform, inconspicuous, and must not interfere with other tasks (Scott and Haugland, 2001). One option for controlling an FES system is through extraction of electromyographic signals from muscles above the level of injury that retain voluntary control (Kilgore et al., 2008). In addition, muscles below the level of injury have been identified as a potential source of command signals (Moss et al., 2011). In subjects with SCIs classified as motor complete, significant muscle activity was identified in 89% of muscles below the level of injury. Another common approach uses a joint angle transducer, which can be positioned across the shoulder or wrist joint (Peckham and Knutson, 2005). The user can control the stimulation by making movements at the joint by activating muscles still under voluntary control.

Although these methods have been used for more than 30 years, they do have some limitations. In people with cervical level SCIs, there are often few muscles left under voluntary control above the level of injury. Mouth or head movements should not be used for control because most people want to be able to eat without help from a care-giver. A more natural control signal would allow people to trigger FES systems in an inconspicuous way that would not interfere with the task being performed. Recently, signals from the brain have been targeted for use as control signals in FES systems.

8.4 Functional outcome measures

Within the research and clinical communities, there is no general consensus regarding assessment of hand function in people with cervical level injuries

152

because of the difficulty of identifying valid, reliable, and sensitive measures (Bryden et al., 2000, 2004). When assessing hand function in people using a neuroprosthetic, matters become even more complicated. The tests that have been most commonly used in previous studies include the Jebsen–Taylor Test of Hand Function, the Sollerman Hand Function Test, the Grasp and Release Test (GRT), and the Activity of Daily Living Abilities Test (ADLAT).

The Jebsen–Taylor Hand Function Test is a quantitative assessment of grasp and release and object manipulation and is commonly used to assess hand function following interventions in people with SCI, but less commonly used during FES neuroprosthetic use (Jebsen et al., 1969; Beekhuizen and Field-Fote, 2005, 2008). The subject completes seven subtests, following standardized instructions, while being timed. The tasks include (1) writing a short sentence; (2) simulated page turning; (3) lifting small, common objects (bottle caps, pennies, paper clips); (4) simulated feeding; (5) stacking checkers; (6) lifting large, light objects (empty cans); and (7) lifting large, heavy objects (full cans).

The Sollerman Hand Function Test consists of 20 activities of daily living that require common hand-gripping strategies, such as using a key, picking up coins, using a phone, and pouring water from a jar (Sollerman and Ejeskär, 1995). The test was designed with tetraplegic patients in mind, and thus it may reflect the needs of the SCI population better than other hand function tests. However, most of the ADLs performed during the Sollerman test require movements from muscles that are proximal to the hand, which can confound the interpretation of the results.

The GRT measures pinch strength, grasp strength, and hand function in people with cervical SCIs (Wuolle et al., 1994; Mulcahey et al., 2004). The subject is asked to pick up and release five different objects of various weights and sizes using only one hand, and pinching ability is evaluated by grabbing a fork handle and stabbing food. One of the benefits of this test is that the objects do not need to be moved across the body, from one side to the other. Because of this, the GRT more accurately reflects hand function alone, rather than more proximal muscle function (Bryden et al., 2000, 2004; Kilgore et al., 2008). The GRT has been used in previous studies to assess hand function after tendon transfers and during use of FES neuroprosthetics (Peckham et al., 2001; Mulcahey et al., 2004).

The ADLAT quantifies changes in hand function and is appropriate for assessing performance in people with cervical SCIs (Stroh and Van Doren, 1994). The ADLAT consists of six activities, including eating with a fork, drinking from a glass, writing with a pen, dialing a phone, using a CD, and brushing the teeth. The test accounts for subject preference, assistance required, and quality of movement. The ADLAT has been used to measure performance in hand function in people with tetraplegia while using an FES neuroprosthetic for hand opening and closing (Bryden et al., 2000, 2004).

8.5 An exemplar of closed-loop neuroprosthetic control of FES

Several groups have recently developed BCIs combined with rehabilitation to produce new combined therapies (Daly et al., 2009; Ang et al., 2010; Broetz et al., 2010; Várkuti et al., 2013). However, one key challenge in these approaches that has not been addressed is the relatively rapid change in the motor cortex that occurs during rehabilitation (Cramer et al., 2007). Because the motor cortex is the primary input to the neuroprosthetic, changes in the motor cortex could affect the neuroprosthetic's performance. In traditional neuroprosthetic approaches, every day that the subject returns to the clinic for therapy the system must be recalibrated to initialize to high performance. This approach is not well-suited to continuous neuroprosthetic rehabilitation use over long durations spanning from days to years. An alternative to this approach is to maintain the system's performance for the user throughout rehabilitation with an adaptive BCI.

In this section, an electroencephalogram (EEG) BCI system via reinforcement learning decoding is used to illustrate how to control a hand grasp/open functional electrical stimulation (FES) device. In an experimental test bed for augmenting rehabilitation with a BCI, the system architecture is validated and compared in a closed-loop environment between controls and subjects living with SCI. The results show how a rehabilitative BCI can continuously adapt to improve performance over four sessions spanning 1 week of use and without daily initialization.

8.5.1 Study participants

The system design function is demonstrated and compared between a control subject and a subject with chronic SCI. The inclusion criteria for recruiting subjects with SCI includes: chronic injury (longer than 1 year), no denervation of target muscles, and C5 or C6 level motor complete injury classified by the ASIA standards. Both subjects in this example were 30-year-old males. The subject with SCI was injured playing football, and his injury (duration = 15 years) was classified by ASIA standards as C3 (single neurological and bilateral motor levels) incomplete (ASIA B), although motor scores of 5 (normal function) were attained at the C5 level bilaterally, with scores of 5 (right) and 3 (left) at the C6 level. All motor scores below C6 level were zero. The subjects had no history of other serious medical issues.

8.5.2 Behavioral function

Hand grasp/open function is chosen as the behavioral function, in which the goal of the task is to enable direct brain actuation of hand closing and opening. To conduct the test, the subjects sit facing a display with their right forearm resting on a table as shown in Figure 8.4a. After a fixation cross is shown on the display for

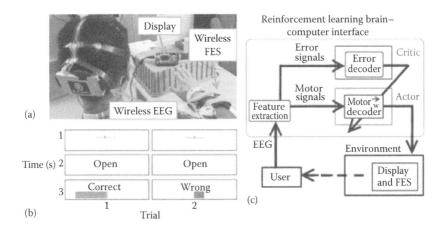

Figure 8.4 (a) Experiment setup overview, visible are the EEG headset, display, and FES. (b) For each trial during the experimental task, the display showed a fixation cross, followed by a cue for "open" or "closed" for 1 second, and then feedback of "correct" or "wrong" for 1 second. A magnitude plot also showed the unthresholded output of the motor potentials decoder. (c) Actor–critic RL BCI architecture. The actor decodes motor potentials and outputs an action. The critic detects an ErrP and provides feedback to the actor. The actor uses feedback from the critic to adapt to the user.

3 seconds to minimize eye movements, cues of "open" or "closed" are presented for 1 second, which instructs the person to either open or close his hand. Visual feedback of "correct" or "wrong" is then shown for 1 second, along with a corresponding plot of the unthresholded output of the system as shown in Figure 8.4a (row 3).

Four closed-loop sessions are performed and consist of 300 trials during the first session, 450 trials each during the second and third sessions, and 300 trials during the fourth session. Time between sessions was varied to test the adaptation of the network with 2 days between the first and the second sessions, 4 days between the second and the third sessions, and 1 day between the third and the fourth sessions. During closed-looped sessions, when the output of the adaptive BCI was determined to be "open," FES was delivered to the hand muscles of the SCI subject. No FES was delivered for trials when the output of the adaptive BCI was "closed." All trials were used in the analysis.

8.5.3 Neural data acquisition

A wireless nine-channel EEG system (256 Hz sampling rate, 16-bit resolution, ×10 headset; Advanced Brain Monitoring, Carlsbad, CA) was fitted to the subject's head (Figure 8.4a). Electrodes (Fz, F3, F4, Cz, C3, C4, POz, P3, P4) were arranged according to the International 10–20 system standards. Foam

155

sensors attached to the sensor sites on the headstrips were saturated with Synapse (Kustomer Kinetics, Arcadia, CA) conductive electrode paste and the corresponding sites on the head were abraded and cleaned before placing the sensors on the scalp. Electrode impedances were tested before and after each experimental session using the manufacturer-provided software. Impedances are generally found to be less than 40 kΩ. Sensors were readjusted on the scalp for proper placement if the electrodes had high impedance values.

ErrPs were recorded from the Cz electrode and the motor potentials for the intent to open or close the hand were recorded from the C3 electrode (Qin et al., 2005; Ferrez and Millan, 2008). For error-related potentials (ErrPs), EEG generated from 0.15 to 0.70 seconds after display of feedback ("correct" or "wrong") was used. For motor potentials, EEG generated between 0.15 and 1.0 seconds after the display of cues ("open" or "closed") was used. The EEG is transformed into the frequency domain using the fast Fourier transform (FFT) to obtain a power spectral density (PSD) of 1 Hz resolution. Frequencies of 1 to 50 Hz are used for the motor potential decoder and frequencies of 1 to 12 Hz are used for the ErrP decoder. The inputs are normalized PSD z-scores (LeCun et al., 1998). The z-scores of the PSD are created by subtracting the mean of previous trials at each frequency and dividing by the standard deviation of previous trials for that frequency.

8.5.4 Muscle stimulation

A neuroprosthetic wrist-hand orthosis (NESS H200, Bioness Inc., Valencia, CA), as shown in Figure 8.4a, was fitted to the right hand of the subject. FES is delivered to the extensor (extensor digitorum communis and extensor pollicis brevis) and flexor (flexor pollicis longus and thenar) muscle groups alternately to produce opening and closing movements of the fingers and hand. Stimulation intensity was set by holding the pulse duration (300 µs) and frequency (35 Hz) constant, while slowly increasing the current amplitude. Once a maximal muscle contraction was attained (increases in current intensity do not produce additional muscle contraction), the current amplitude was increased by an additional 25% to maintain consistent muscle contractions throughout the experiment.

8.5.5 Actor–critic reinforcement learning architecture

The adaptive BCI decoder shown here is based on an actor–critic reinforcement learning (RL) architecture as shown in Figure 8.4c (Mahmoudi and Sanchez, 2011). The actor decodes motor potentials from the user to determine the user's intent to open or close the hand. The critic provides feedback to the actor by detecting ErrPs generated by the user (Falkenstein et al., 2000). The actor–critic RL algorithm is a semisupervised machine learning algorithm that optimizes the actor's decoding of the user's motor potentials based on feedback from the critic (Sutton and Barto, 1998).

The actor is parameterized by a three-layer fully connected feedforward neural network. The hidden and output nodes of the neural network perform a weighted sum on their inputs. The weighted sum at each node is passed through a hyperbolic tangent function with an output in the range of -1 to 1. The weights between the actor's nodes are initialized randomly and then updated after each trial based on feedback. The actor's weights update can be expressed as $\Delta w_{ij} = \gamma f(x_i(p_j - x_j)) + \gamma(1 - f)(x_i(1 - p_j - x_j))$. Here, w_{ij} is the weight connecting nodes i and j, γ is the learning rate, p_j is a sign function of output x_j (positive values become $+1$ and negative values become -1), and f is feedback from the critic. The weight update equation is based on Hebbian style learning (Mahmoudi and Sanchez, 2011; Pohlmeyer et al., 2012). The critic provides the feedback by decoding the user's EEG to determine if an ErrP was generated. If an ErrP is detected, a feedback of -1 is provided to the network for adaptation. If not, a feedback value of 1 is given. The functional mapping between neural activity and behavior in the actor is constructed using the weight update equation above.

8.5.6 Adaptive BCI usage

Adaptive BCI usage is broken down into several intermediate steps, as shown in Figure 8.5. Representative ErrPs are collected in the preliminary session and used to develop the critic through supervised learning (Prechelt, 1998). Once the critic is created, the weights of the actor are initialized to random initial values and trained through RL and feedback from the critic. After the first closed-loop session, in which the weights are initialized to random values, all subsequent closed-loop sessions use the weights from the previous session with no offline adjustments.

8.5.7 Critic as error potential classifier

The error potential classifier "critic" detects ErrPs in the user's EEG to determine if the user perceived that an error occurred. The critic then provides binary feedback, -1 or 1, to the actor. The input to the error potential classifier is the normalized PSD from 1 to 12 Hz in 1-Hz bins computed on the 0.15

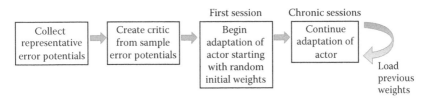

Figure 8.5 Flowchart shows the preliminary steps of the experiment and how the final step can be repeated.

157

to 0.70 seconds of EEG data after the actor's output (action) is shown on the display.

The error potential classifier in the critic is a three-layer neural network with 12 input nodes for the 1 to 12 Hz in 1-Hz bins and five hidden nodes. Representative ErrPs are collected in the 120 trials of the preliminary session and are randomly assigned to either a training set or a test set, for approximately 60 trials each. The training set is used to optimize the weights of the critic with supervised learning. The weights with the best classification accuracy are used for closed-loop sessions. To test the critic training during the preliminary data collection, 10-fold cross-validation is performed.

8.6 Closed-loop trials

Figures 8.6 and 8.7 show representative trials from the closed-loop experiments and give insight into how the system processes the EEG to create features for the classifiers. The first row of Figure 8.6 shows the filtered (1–50 Hz) EEG from the C3 electrode for the 0.15 to 1.0 seconds after the cue is presented. The second row shows the PSD computed from the raw EEG. The z-scores of the PSD are shown in the third row as inputs to the actor. The first column shows the filtered EEG and processing after an "open" cue. Similarly, the second column shows the filtered EEG and processing after a cue of "closed" was shown. The features for the cue of "closed" correspond to lower power, in general, than the features of the cue for "open"; in the sample trial of the SCI subject, 44 of the 1 Hz bins have lower power for the "closed" cue.

A similar process is used for inputs to the critic. The first row of Figure 8.7 shows the filtered 1 to 12 Hz EEG from the Cz electrode for the 0.15 to 0.70 seconds after the feedback is shown. The PSD of the raw EEG was computed from the Cz electrode, shown in the second row. Finally, the inputs to the critic are shown in the third row as z-scores of the PSD from the Cz electrode. The first column shows the filtered EEG and processing after the feedback of "correct" was presented. The second column shows the filtered EEG and processing for feedback of "error." Notice that the error potential has a biphasic shape characteristic of this neural oscillation. The features for feedback of "correct" correspond to lower power, in general, compared to features of "error"; in the sample trial for the SCI subject, all 1 Hz bins except 1, 8, 11, and 12 Hz. Figure 8.8 shows the ErrPs generated by the users, the average of error trials minus the average of correct trials. The ErrPs collected from the users are similar to published results (Ferrez and Millan, 2008).

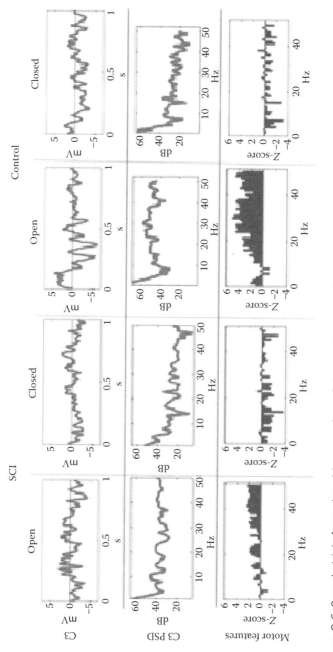

Figure 8.6 Sample trials from closed-loop sessions. Columns show samples for cues and feedback of "open" and "closed" for both the SCI and control subject. Rows show raw EEG from electrode C3, PSD, and features.

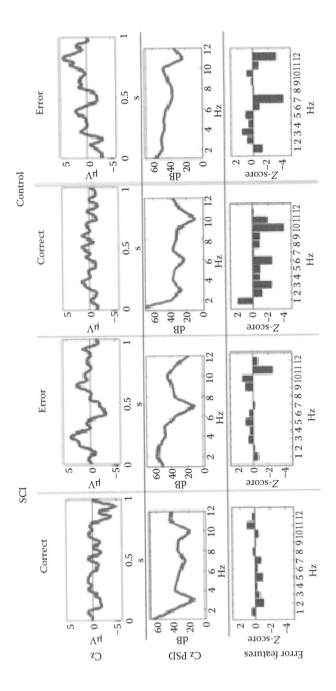

Figure 8.7 Sample trials from closed-loop sessions. Columns show samples for cues and feedback of "correct" and "error" for both the SCI and control subject. Rows show raw EEG from electrode Cz, PSD, and features.

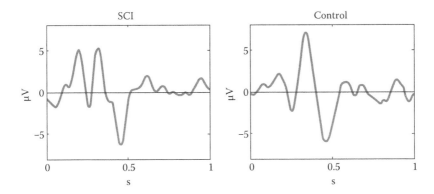

Figure 8.8 Averaged trials and error-minus-correct for the SCI and control subjects.

8.6.1 Performance over time

Figure 8.9 shows the overall performance of the actor in classifying motor potentials across four sessions for control and SCI subjects. The classification accuracy starts below 50% (chance level) for the SCI subject because of the random initial values of the actor's weights. The performance of the actor improves as the actor's weights adapt to feedback from the critic through RL. Over time, the actor's performance approaches the classification accuracy of the critic. Changes in weight values become smaller after the first two sessions; however, changes in weight values continue throughout the 1500 trials. The actor makes fewer mistakes during the last session than the first, as the actor adapts and learns the user's motor potentials based on feedback from the critic.

8.6.2 Comparison of performance across subjects

The overall performance of both subjects across sessions is shown in Figure 8.10. The subjects had comparable performance, above chance level (50%) starting at the end of the first session. The performance of the control subject was slightly higher than that of the SCI subject during the first session. This performance difference can be explained by the random initial weight values of the actor more closely matching the desired weight values by chance. The overall performance of the SCI subject was only slightly lower (by 0.9%) than the control subject. The system also had lower accuracy for detecting the SCI subject's ErrPs (64.2%) than for the control subject (68.8%). This lower performance in detecting the SCI subject's ErrPs could explain the lower overall performance of the SCI subject compared with the control subject. Importantly, the performance of the critic had a small standard deviation (3.6%) for the SCI subject.

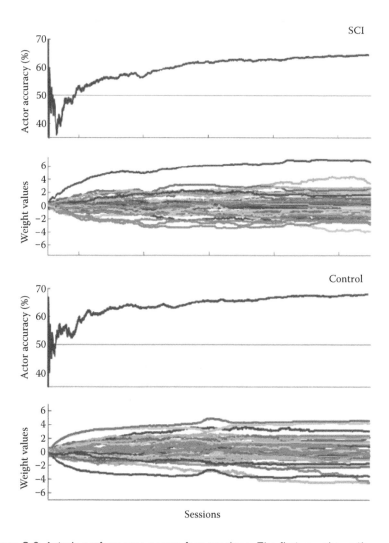

Figure 8.9 Actor's performance across four sessions. The first row shows the actor's cumulative classification accuracy and the second row shows the actor's weights adapting for the SCI subject. The third row shows the actor's cumulative classification accuracy and the fourth row shows the actor's weights adapting for the control subject. Final cumulative classification accuracies were significantly above chance (50%) for both subjects ($p < 0.001$, one-sided t-test).

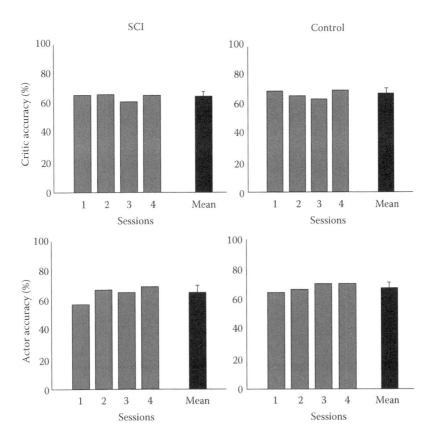

Figure 8.10 The first row shows the accuracy of the critic for both the SCI and control subjects. The second row shows the accuracy of the actor. Accuracy for each day is shown in gray. Mean accuracy across days is shown in black.

8.7 Conclusion

This chapter developed the concepts for designing an EEG-based BCI system using RL to control a hand grasp FES device for augmenting rehabilitation with a BCI. The system used RL to determine the mapping of motor potentials to intended actions based on user-generated ErrPs. The BCI continued to adapt to the users throughout the experiment and did not require any offline training after the first session. The ability to adapt to the user without daily initialization could be beneficial in a rehabilitation setting. Cortical reorganization from the rehabilitation could change the user's motor potentials, increasing the need for daily adjustments to the system. After an SCI, the brain experiences measurable maladaptive brain reorganization from disuse (Green et al., 1998; Cramer et al., 2005;

Hoffman and Field-Fote, 2006; Kokotilo et al., 2009). The ability to rehabilitate the motor cortex by motor imagery alone is important in the context of BCI-augmented rehabilitation because motor imagery is often used to control BCIs. Notably, motor imagery has been used to control hand grasp FES in BCI systems (Pfurtscheller et al., 2003; Müller-Putz et al., 2005). The combination of motor imagery and BCI-controlled FES has been shown to rehabilitate finger extension in a stroke subject (Daly et al., 2009). This improvement occurred with only three sessions a week for 3 weeks. Compared with other interventions such as constraint-induced movement therapy, this is a limited amount of time participating in rehabilitation (Liepert et al., 2000). By using an adaptive BCI, the subjects can participate in rehabilitation over a longer period similar to constraint-induced movement therapy without needing to stop the rehabilitation to recalibrate the system. Maintaining continuity in the performance over time is a critical aspect in the rehabilitation process. The user is able to pick up from the last level of progress achieved from the previous session.

The concepts presented in this chapter also open the possibility for the subjects to take the system home and use it continuously. This is due to not only the continuous RL that does not require calibration by a scientist but also to the design, which uses the commercial Bioness H200 and an easy-to-use wireless Advanced Brain Monitoring EEG system. Several additional results are also applicable to the use of the system during rehabilitation. The decoder continued to adapt even in later trials, so the system can be expected to continue to adapt to the user in future trials and during rehabilitation. The performance of the system increased above chance during the first day and continued to show improvement in later trials, both factors in maintaining user motivation and engagement.

Exercises

1. Perform a hypothetical ASIA exam for a person who has experienced a spinal cord injury from a motorcycle accident.
2. In the example BCI for rehabilitation shown in this chapter, the decoder was designed for four sessions. Design an adaptive decoder that would operate for 1 month of use. What aspects of the actor or critic would need to be changed to support the activities of daily living that occur over a month? How would the system be used over that time? Include details about the setup, deployment, and evaluation of function.
3. True or false. After injury, the brains of people living with spinal cord injury experience maladaptive brain reorganization in which they cannot generate kinematic training signals for neural decoders.

Chapter 9 Design of implantable neural interface systems

The medical and social community was not ready for electrostimulation: Hyman's device was roundly dismissed as "gadgetry" that interfered with natural events at best and the work of the devil at worst.

Oscar Aquilina
A Brief History of Cardiac Pacing, 2006

Learning objectives

- Define the major signal acquisition, processing, and actuation components of a neuroprosthetic system.
- Design and optimize the architecture of a fully implantable neural interface system for any application.
- Specify the trade-offs that affect the performance of a fully implantable neuroprosthetic system.

9.1 Introduction

The design of fully implantable neural interface systems is a highly multidisciplinary effort that combines fundamental neuroscience, neural engineering, electrical engineering, and clinical care. At its core, design principles focus on the development of direct interfaces to the nervous system for delivering precise, knowledge-driven approaches to restoring or augmenting function of the brain. Before the advent of neural interface systems, clinical approaches for treating the dysfunction or injury to the nervous system were not precise; consequently, physicians were limited in their ability to directly interact with specific neurons or neural networks that were contributing to the problem, as shown in Figure 9.1. For example, in the case of neuropsychiatric illness, the traditional standard of care is to present the illness to a psychiatrist who can prescribe behavioral therapy or pharmaceuticals as a method of therapy. However, both have their limitations in terms of precision. Behavioral therapy is highly indirect because modifications

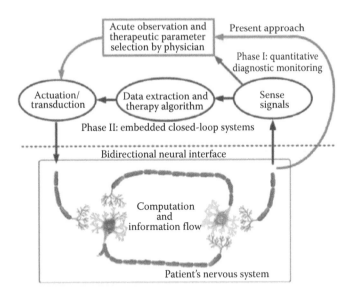

Figure 9.1 Architecture for general-purpose neural interface system. Signal sensing, data processing, and direct actuation provide precision compared with traditional clinical approaches (pharmaceuticals) for treatment of the nervous systems. (Adapted from Rouse, A. G., S. R. Stanslaski, P. Cong, R. M. Jensen, P. Afshar, D. Ullestad, R. Gupta, G. F. Molnar, D. W. Moran, and T. J. Denison. A chronic generalized bidirectional brain–machine interface. *J. Neural Eng.* 8(3):036018. doi: papers2:// publication/doi/10.1088/1741-2560/8/3/036018, 2011.)

of actions are used to influence the signaling in the brain. The mechanisms of influence of how behavioral therapy affects specific networks are not known. In contrast, pharmaceuticals are more direct in their influence on the nervous systems; however, that influence is very generic, and thus it influences many circuits of the brain simultaneously with very limited ability to target.

The fundamental advantage of neural interface systems compared with traditional clinical therapies is that one can design for specificity and direct interaction with the nervous system. To do so, the designer must first determine when and how the disease is affecting the patient. This includes identifying the critical circuits and subcircuits that are involved in the neural information processing of the disease process. Once those circuits are identified, the details of the signaling involved in the neural information processing must be identified, as shown in "Phase I" of Figure 9.1. Often called the "features" of neural signaling, selection of the appropriate level of abstraction (spikes, local field potentials, electrocorticograms, electroencephalograms, etc.) must be considered in terms of their accessibility and the technical feasibility of acquiring those signals. Once the appropriate signals are identified, the design of closed-loop systems can be enabled to provide the ability

166

Table 9.1 Example Neural Interface Types and Their Applications

Neural Interface	Indication	Number of Implants
Deep brain stimulator	Parkinson's disease, distonia	>100,000
Spinal cord stimulator	Chronic pain, incontinence	>140,000
Cardiac pacemaker	Replacement of natural heart pacemaker	>350,000 implants per year
Cochlear implant	Restoration of hearing	>100,000

to directly sense, process, and actuate the nervous systems in real-time, as shown in "Phase II" of Figure 9.1. The ability to chronically sense, process, and tele-meter signals from the nervous system can help address disorders often believed to be intractable because direct observations and knowledge-driven interactions with the nervous system provide new principles of therapy. This process can lead to improved monitoring of disease progression and therapy efficacy in applica-tions. In addition, precision interfacing with the nervous systems leads to an under-standing of neural dynamics, which guides the development of embedded sensors and chronic signal classifiers that help to optimize and deliver therapy in real time.

In addition to the fundamental design of neural interface systems, important con-siderations include the substantial development and clinical trial costs of devel-oping such a device for human use. The designer must incorporate the practical issues that include chronic deployment such as implantation technique, reliability of components, chronic signal integrity, and power consumption. The referenc-ing of predicate devices can play a critical role in the design strategy of next-generation interfaces. As shown in Table 9.1, there are many neural interface systems that contain components related to the general conceptual architecture presented in Figure 9.1. These include deep brain stimulators for Parkinson's dis-ease, spinal cord stimulators for chronic pain and incontinence, cardiac pacemak-ers, and cochlear implants for restoration of hearing (Stieglitz, 2010). Combined, these systems have a set of users on the order of one million people. Guidance can be obtained from the clinical translation of these devices to that of novel architec-tures and the rigorous validation and reliable analysis needed for a diverse set of real-world environments.

9.2 Design

Necessary requirements for the chronic use of neuroprosthetics include small size, low power, and biocompatibility to enable the development of reliable, min-iature, subcutaneous implants capable of transducing, amplifying, and wirelessly transmitting neuronal activity. Tethered prototype neural interface systems, such as those that have been designed since the early 2000s, limit the movement of

167

the subject, increase the potential for infection, and introduce a number of artifacts in the recordings. These drawbacks motivate the use of wireless implantable recording devices. Power, area, and bandwidth are active constraints that limit the design of wireless neural interface systems. The resolution of these difficulties requires the use of new fabrication techniques, materials, and architectures for the implanted electrodes as well as novel computational architectures that can exploit the neural signal characteristics.

The challenge of building a fully implantable wireless neural interface system has been undertaken by several academic and industrial institutions (Wise et al., 2003, 2008; Patrick et al., 2009, 2010; Konrad and Shanks, 2010; Nurmikko et al., 2010; Rapoport et al., 2012; Stanslaski et al., 2012; Zhang et al., 2012; Borton et al., 2013; Chen et al., 2013; Poppendieck et al., 2014; Ryapolova-Webb et al., 2014). Each system has its respective advantages and disadvantages depending on the site of implantation, targeted signal acquisition, power/bandwidth trade-offs, implant size, and fabrication technology. Systems consisting of silicon shank technology have the advantage of system integration of a well-established lithographic technique. One trade-off for this approach is the rigidity of the substrate, which may induce biotic–abiotic interface issues (Streit et al., 2012). In addition, silicon substrates are brittle compared with metals and have greater potential for fracture. In addition to the substrate, the physical layout of the electronics plays a significant role in functionality. The design developed by Harrison et al. (2009) achieves integration by stacking the electronics on top of a "bed-of-nails" substrate. Because the electronics are located on top of the probe, tissue heating and extra momentum transfer to the tissue could induce biocompatibility problems. In contrast, by distributing the electronics in a way that is similar to devices being developed for deep brain stimulation (DBS), power, bandwidth, and size constraints are overcome using chest-implanted devices with wires routed to the cortical implant. The solution to overcome this limitation is to separate the back-end electronics from the recording site. These and other major design considerations for neural interface systems are investigated in the next sections.

9.2.1 Effect of signal choice

The selection of signals to interface with in the nervous system ultimately drives the specifications for chronic recording, information coding, and overall system power, as shown in Figure 9.2 (Rouse et al., 2011). With current electrophysiological approaches, spike recording provides the highest spatiotemporal resolution into the nervous system because it directly interacts with the output action potentials of single neurons. Spike recording is challenging because of the highly dynamic biotic–abiotic interactions that occur in the microenvironment between the microelectrode and the neuron itself. Acquisition of neural signaling is not guaranteed

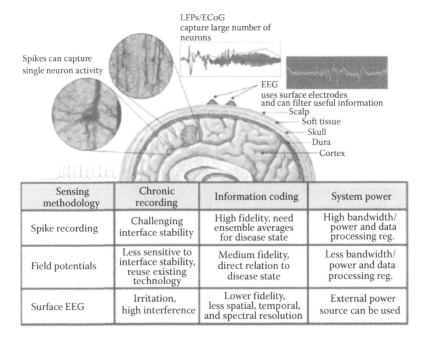

Sensing methodology	Chronic recording	Information coding	System power
Spike recording	Challenging interface stability	High fidelity, need ensemble averages for disease state	High bandwidth/ power and data processing reg.
Field potentials	Less sensitive to interface stability, reuse existing technology	Medium fidelity, direct relation to disease state	Less bandwidth/ power and data processing reg.
Surface EEG	Irritation, high interference	Lower fidelity, less spatial, temporal, and spectral resolution	External power source can be used

Figure 9.2 The selection of signal sensing that drives specifications for chronic recording, information coding, and overall system power. (Adapted from Rouse, A. G., S. R. Stanslaski, P. Cong, R. M. Jensen, P. Afshar, D. Ullestad, R. Gupta, G. F. Molnar, D. W. Moran, and T. J. Denison. A chronic generalized bi-directional brain–machine interface. *J. Neural Eng.* 8(3):036018. doi: papers2://publication/doi/10.1088/1741-2560/8/3/036018, 2011.)

because of micromovement and degradation of the integrity of the electrode interface. Although single neuron action potentials have high fidelity, a representation of the physiological state must be inferred because of the mismatch between the single neuron element and the overall nervous system hierarchy, which contains considerably more neurons. However, if enough neurons are simultaneously sampled, this mismatch can be offset by analysis of the population. The simultaneous recording of even a few neurons introduces significant bandwidth, power, and processing requirements. Because single neuron action potentials contain spectral information in the 500-Hz to 8-kHz bandwidth, sample rates for conventional analog-to-digital converters (ADCs) are in the 20 kHz range. The total bandwidth quickly scales by the number of neurons, sample rate, and the ADC sample resolution. Processing requirements are also exacerbated by the need to perform spike sorting to localize each individual neuron (Lewicki, 1998). Depending on the methodology chosen, the complexity can scale from the more efficient template matching to the more demanding principal components analysis (PCA) methods. By moving up a level of the hierarchy in the organization of the nervous system to the local field potentials (LFP), multiple constraints can be relaxed. Compared with single neuron activity, LFPs reside in the 0.5-Hz to 500-kHz band of the spectrum and contain elements of dendritic activity in the volume of tissue surrounding a population of neurons. Because LFPs are captured in the vicinity of multiple neurons, their representation collectively is more indicative of the overall population state. In addition, because the precise relationship between the sensing electrode and the neuron is not as critical as in the spike recording case, this methodology is not as sensitive to the interface stability and can be reliably recorded using a variety of microscale and mesoscale electrodes ranging in size from 10 μm to 4 mm in diameter. Likewise, because there is less bandwidth and no spike sorting required, this methodology provides a good balance in system power and data processing. Lastly, the electroencephalogram (EEG) is the least invasive and most accessible neural interface; however, there are multiple issues with sensing neuronal activity through the cerebral spinal fluid, dura, skull, and scalp. These interfaces are subject to interference from muscle and ambient activity, which overlaps with their spectral content in the 0.5- to 80-Hz range. Although the EEG temporal resolution still remains high, the spatial resolution is greatly diminished and is in the order of 1 to 2 cm, and thus makes inferring specific cortical representation difficult. With EEG acquisition being all externalized, system power is not a big factor because all of the associated electronics can also be deployed external to the body.

9.2.2 Three primary constraints

Once the appropriate signal for the neural interface application is chosen, the designer must then balance size, power, and functionality for the hardware design, as shown in Figure 9.3 (Sanchez et al., 2008; Bashirullah, 2010). The implications of restricting the size of the instrumentation hardware is particularly limiting on the overall functional specifications of the device because it affects the signal-processing efficiency and capacity across the entire system. For instance, a battery-operated neural interface instrumentation hardware with stringent size restrictions (~1 cm^2) is functionally limited by the size of the battery and hence the available capacity, peak power handling capability, and overall lifetime. In addition, the size restrictions also limit the attainable radiation efficiencies of electrically small antennas that are often used in biomedical implants to decrease the complexity, power, and operating frequency of the radiofrequency (RF) transmitter and minimize signal loss because of skin/tissue attenuation. A device implemented entirely with commercially available off-the-shelf components cannot achieve the levels of integration as in custom chip solutions and may incur additional power losses from driving off-chip board level components and from communication overhead between modules. On the other hand, integrated solutions based on conventional signal processing can achieve high levels of functionality with small overall device footprint, but generally at the expense of increased

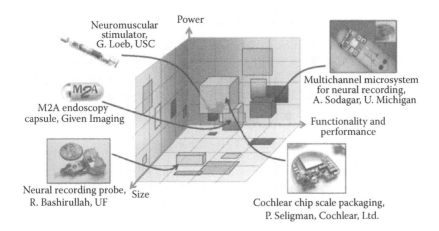

Figure 9.3 The relationship between size, power, and functionality for neural interface systems. (Adapted from Bashirullah, R. Wireless implants. *IEEE Microwave Mag.* 11(7). doi: 10.1109/MMM.2010.938579, 2010; Sanchez, J. C., J. C. Principe, T. T. Nishida, R. Bashirullah, J. G. Harris, and J. A. B. Fortes. Technology and signal processing for brain–machine interfaces. *IEEE Signal Process Mag.* 25(1):29–40. doi: 10.1109/MSP.2007.909525, 2008.)

power dissipation overhead associated with processing and wireless transmission of high bandwidth and high-resolution signals.

9.2.3 Implant location and form factor

Many of the design constraints of neural interface systems can be mitigated with careful choice of the implant location. This is achieved with appropriate routing of the leads, placement of the battery (depending on size), and choice of where onboard processing is conducted. To illustrate the importance of implant location (Figure 9.4), consider the following: systems fully implanted in the cranium have very rigorous size constraints where the battery, processor, and wiring must all be compactly restrained to a few square centimeters of surface area. In addition, heating of tissue is a concern because of the onboard processors being located so close to the brain. A system implanted in the chest near the pectoral muscle can provide a balance to the size constraint problem because larger components such as the battery can be distally connected where there is more surface area. The trade-off is that the battery wiring must be wired through the neck to the implant site. One emerging technology to mitigate this problem is the design of body area networks (Maskooki et al., 2011). Systems that distribute cranial and distal implants can achieve the best of both locations. Low-power local processing can be achieved in the cranium to condition signals, which could then be routed to the distal site where complex processing and larger batteries could be deployed.

171

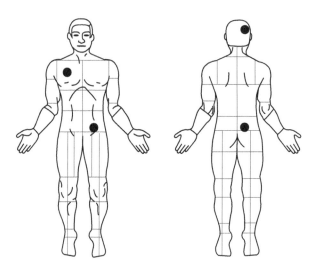

Figure 9.4 Common implant sites for neural interface systems.

9.2.4 Fabrication and construction

To encompass a systems-based hardware development component in order to create safe and effective sensing and stimulation systems, the choice of neural signal, battery, onboard processor, telemetry, and implant location must be integrated. The design of such integrated systems could be based on currently available commercial technology or take on a new form factor. Additionally, the task of building the device must fall within parameters accepted by both the Food and Drug Administration and end users from the clinical and patient communities. Figure 9.5 illustrates the six components that comprise the fabrication and overall construction of the device (Stieglitz, 2010). They include the product design and intended use, which could be for sensory, motor, cognitive, or neuropsychiatric applications. The selection of the specific internal electronic components that are capable of delivering the desired functionality must be arranged to fit within the form factor most amenable to the application and implant location. A critical step in the manufacturing for large-scale application includes the choice of assembling techniques needed to connect the relevant sensing, stimulation, computational, and communication components to the circuit board substrate. It is often the case that there are multiple modules that are needed to support power, computation, and telemetry functionality. Given the small form factor for most neural interface implants, the interconnects for assembling multiple chips becomes a challenging design consideration because of the availability of commercial technology that can meet the bandwidth and channel counts commonly associated with interfacing with neural signals. Ultimately, the overall footprint of the electronics package is governed by the housing and external interconnects needed to protect the

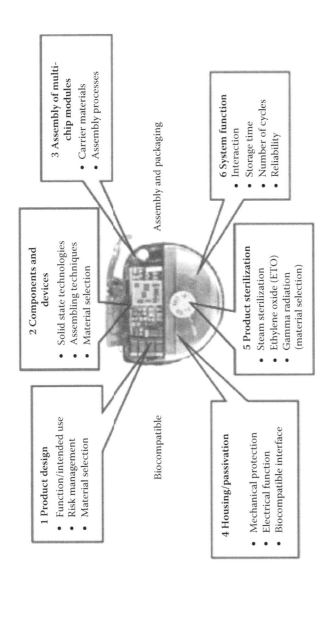

1 Product design
- Function/intended use
- Risk management
- Material selection

2 Components and devices
- Solid state technologies
- Assembling techniques
- Material selection

3 Assembly of multi-chip modules
- Carrier materials
- Assembly processes

4 Housing/passivation
- Mechanical protection
- Electrical function
- Biocompatible interface

5 Product sterilization
- Steam sterilization
- Ethylene oxide (ETO)
- Gamma radiation (material selection)

6 System function
- Interaction
- Storage time
- Number of cycles
- Reliability

Assembly and packaging

Biocompatible

Figure 9.5 Common implant sites for neural interface systems. (Adapted from Stieglitz, T. Manufacturing, assembling and packaging of miniaturized neural implants. *Microsyst. Technol.* 16(5):723–34, 2010.)

internal circuitry from mechanical and biological sources of interference. The housing must be robust to product sterilization. Steam sterilization, which uses temperature and pressure as a part of its protocol, produces the most significant physical effect. Other approaches such as the use of ethylene oxide are often more amenable for delicate electronic or microelectrode arrays. Lastly, the overall system integration must serve the desired function and application of the device. If data are stored onboard, the capacity should be sufficient (i.e., 24-hour recording) to provide a benefit to the user or clinician. In addition, the desired memory architecture must fit within the physical constraints of the implantable package. Likewise, for the onboard processor, the computational complexity of the neural signal-processing algorithms needs to fall within available technology that can be deployed in the package and serve the ultimate needs of the intended use of the device. Because the relationship between implant size is related to the complexity of the overall system being developed, there is a spectrum of options for manufacturing, as shown in Figure 9.6. Highly complex devices that require small packages are well suited for micromachining approaches to fabrication. In contrast, low complexity devices that are large are better suited for precision mechanics. As in all systems designs, the overall approach is not limited to these extremes and can also consist of a blended form of these fabrication methodologies.

9.2.5 Hermetic packaging

To endure a lifetime of implantation in the human body, the sensitive electronics within a neural interface system must be protected from biologic environments that consist of water and ions in solution. Although corrosion and degradation of the bulk of the implant may seem to be the most obvious considerations, there are multiple subtle aspects of the design of hermetic packaging that make it a difficult and often underappreciated engineering challenge. Because neural implants function as a bridge between the externalized biological tissues and the internalized processing systems, there are natural biotic–abiotic interfaces that can be pervious to the ionic solutions. These can occur particularly where there are wire bonds,

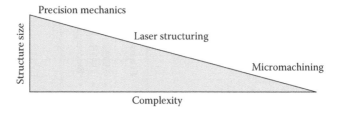

Figure 9.6 Required manufacturing technology for neural interface systems is governed by the relationship between the structure size and complexity of the overall system. (Adapted from Stieglitz, T. Manufacturing, assembling and packaging of miniaturized neural implants. *Microsyst. Technol.* 16(5):723–34, 2010.)

174

connectors, or the joining of two dissimilar components of the system. Hermetic sealing can also occur on a variety of levels spanning from the entire system being hermetically sealed to an overall nonhermetically sealed system that contains hermetically sealed components. The choice depends on the system design, in which bulk-assembled technologies are more amenable to fully hermetic systems whereas distributed interfaces over large surface areas often include a mixture. A variety of materials and housing designs have been developed to contend with the problem of hermetically sealing neural interface systems. On the system level, hermetic packages are often made out of metal and ceramics. Since they have a bulk design, they also contain all electronic components in one assembly. Metals, such as titanium and alumina (Al_2O_3), and ceramics are the materials most commonly used. For devices that include a mixture of hermetic and nonhermetic sealing on the system level, polymers such as silicone rubber and parylene C are the materials of choice. All polymer materials are considered nonhermetic by nature according to the definition of hermeticity, that is, a helium leakage rate of less than 10^{-9} mb/s (Stieglitz, 2010). The longest experience has been gained with silicone rubber and it has been used to encapsulate electronics and their associated wiring for implants that have been in the body for decades. Parylene C is also an attractive material because of its large resistivity (insulator), flexibility, and low permeability (against water, gases and ions), as well as its approval in the United States for chronic human use. In general, the encapsulation material should block the migration of ions and the condensation of water on or around the electrical components.

9.2.6 Considerations for battery recharging and telemetry

A variety of core components and architectures were developed in the early 2000s in response to the specific requirements encountered when acquiring neural signals (Wise et al., 2008). The choices included in each design depend primarily on the selection of neural signals, where they are processed, and which information is transmitted out. As described earlier, if the full bandwidth of neural signaling is sought, this requires the most advanced telemetry system. In contrast, if all neural signals are processed onboard and only actuation is sent out wirelessly, then the least amount of bandwidth is needed and more conventional telemetry can be used. To illustrate this trade-off, three systems are compared for their ability to transmit neural signals. The Michigan implantable 64-channel wireless microsystem consists of a silicon microprobe array, preamplifier, neural data processing unit, and wireless transmission. The amplified neural signal is digitized by the 8-bit conventional synchronous ADCs at a speed of 62.5 kS/s. Because the transmission bandwidth is 2 Mb/s, the microsystem can transmit only the full waveforms on 2 channels or only the spike occurrence and its site of origin on 64 channels. Similar to the silicon shank approach, the Utah multichannel neural recording systems also digitizes signals by synchronous ADCs and can only transmit the full waveform on one channel (Harrison et al., 2009). To efficiently use the bandwidth, the Utah implantable neural recording system performs spike detection and transmits only the spike timings. However, this solution limits the ability to reference action potentials with the background activity because there is no access to the original signal. This bandwidth limitation is avoided by the Brown University approach, which transmits the digitized neural signals through the skin by infrared light using a laser diode (Borton et al., 2013). The infrared approach can achieve a bandwidth of up to 10 Gb/s. However, the infrared telemetry consumes 12 mV, and its high power consumption may prevent it from being used in the proximity of brain tissue.

Despite the variety of approaches for wireless neural interfaces, there are core components that are commonly used, as is illustrated in Figure 9.7. The externalized part of the system typically contains a class E amplifier, which is connected to an externalized low-voltage battery. These are connected to a primary coil, which induces a voltage in a secondary coil on the implant side.

The secondary coil is passively resonated using a capacitor to boost the voltage and then rectified and regulated to provide a clean supply for the on-chip electronics (Bashirullah, 2010).

Wireless powering of implantable neural interface systems can be achieved using low-frequency inductive links, which can penetrate biological tissues. Figure 9.8 (Courtesy of Intertek.com. Available at http://www.intertek.com/uploadedFiles/Intertek Divisions/Commercial_and_Electrical/Media/PDF/Medical_Equipment/EMC -Active-Implantables-Medical-Devices-WP.pdf) shows the relationship between frequency and penetration depth. Here, a 5.8-GHz signal can maximally penetrate 2.6 cm whereas 13.56 MHz can penetrate more than 6 cm. Note that the relative field strengths also vary greatly between the two extremes. As a general rule, the magnetic field strength over the coil axis decreases as a third power of distance.

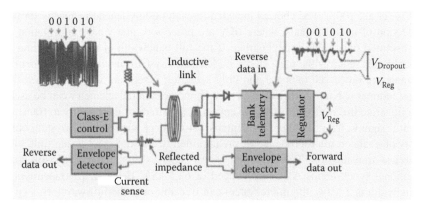

Figure 9.7 Core components of a wireless data and power telemetry system for neural interfacing. (Adapted from Bashirullah, R. Wireless implants. *IEEE Microw Mag* 11(7). doi: 10.1109/MMM.2010.938579, 2010.)

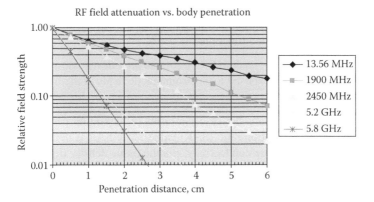

Figure 9.8 Relationship between frequency and tissue penetration depth. (Courtesy of Intertek.com. Available at http://www.intertek.com/uploadedFiles /Intertek/Divisions/Commercial_and_Electrical/Media/PDF/Medical_Equipment /EMC-Active-Implantables-Medical-Devices-WP.pdf.)

Thus, inductive links are only suitable for very small distances. A second complication of inductive interfaces is that the overall coupling strength is dependent on coaxial alignment, coil separation, coil geometry, and angular alignment. Alterations of any of these properties can greatly affect the performance of the inductive link. One important design feature is that, for these systems, the power and data links are commonly separated in frequency and in space such that they can be independently optimized.

Information is transmitted over the channel using three common techniques: on/ off keying, amplitude shift keying, and frequency shift keying. By modulating the state, amplitude, or frequency of the signal, variations in the secondary coil can be created/detected such that binary information can be transferred. For example, in an amplitude shift keying design, the binary symbol 1 is represented by transmitting a fixed-amplitude carrier wave for a specified duration. If the signal value is 1, then the carrier signal is transmitted; otherwise, a signal value of 0 is transmitted.

9.3 Safety

Designers of fully implantable neural interface systems must address safety concerns regarding their devices. This includes not exceeding acceptable and safe levels of tissue heating or exposure to electric/magnetic fields or doses of particles. To minimize RF heating due to tissue absorption, these inductive links are generally operated at less than 10 MHz with typical output power ranging from 10 to 250 mW. Additionally, designers must address the robustness of their device to recording artifacts including motion, nocturnal/diurnal cycles of the brain, and electrical interference generated through simultaneous stimulation and recording. Sophisticated designs are needed to minimize such components of safety while attaining weight, power, and functionality goals.

The IEEE has created the standard C95.1-2005 to provide a guide against harmful effects in human beings exposed to electromagnetic fields in the frequency range from 3 kHz to 300 GHz.

These recommendations are expressed in terms of maximum permission exposures and specific absorption ratio (SAR) values. The SAR is defined as $SAR = \sigma E^2/\rho$, where σ is the tissue conductivity, r is the tissue density (kg/m^3), and E is the root mean square (rms) electric field strength in tissue (V/m). For power transfer through inductive links, the resulting RF heating of the tissue must be kept to less than 1°C to 2°C to avoid cellular damage in sensitive areas such as the brain.

9.3.1 Device interactions

The safety of fully implantable neural interface systems is not limited to just the interactions between the device and the body. Considerations of the potential interactions between the device and the external environments should also be evaluated. In many cases, subjects that have fully implantable neural interface systems are also undergoing extensive medical evaluation. Significant additional risks come in the form of the ferromagnetic case interfering with magnetic resonance imaging fields that can cause serious injury or even death. The subject could also have other medical devices being used on them such as an external defibrillator, which could send large electrical pulses through the neural interface. Electrocautery systems are also often used during surgery for wound maintenance and could interfere with onboard electronics. High radiation sources could also induce electronics damage. Procedures for conditions unrelated to the nervous system such as lithotripsy could present a scenario in which ultrasonic pulses could damage the neural interface system. Lastly, as fully implantable medical devices become more common, it is not unlikely that an individual will have multiple implanted devices and there is a risk of interference among them.

Exercises

1. What are the signal trade-offs in the design of implantable neuromotor systems? Compare and contrast aspects of chronic recording, information coding, and system power.
2. You are tasked to build a neuromotor interface for the following problems: (1) arm paralysis, (2) Parkinson's disease, and (3) restoration of sensory function. Which neural systems would you target for each and why? What are the basic design components for each system?

Chapter 10 Application
Deep brain stimulation for neuropsychiatric disorders

It would appear that we have reached the limits of what it is possible to achieve with computer technology, although one should be careful with such statements, as they tend to sound pretty silly in 5 years.

John von Neumann

Learning objectives

- Understand current and future applications of deep brain stimulation for neuroprosthetics in the cognitive domain.
- Design a neuroprosthetic system for measuring, modeling, and intervening in neuropsychiatric disorders.
- Determine how to validate and enhance the precision of the neurophysiological and functional effects of closed-loop neuroprosthetics for neuropsychiatric disorders.

10.1 Introduction

Deep brain stimulation (DBS) has emerged as a highly efficacious treatment for addressing basal ganglia disorders such as Parkinson disease (PD), essential tremor (ET), and dystonia (Visser-Vandewalle et al., 2004; Benabid, 2010; Eller, 2011; Hu et al., 2011). The benefits of DBS are broad because the therapy can be reversible, adjustable, bilateral, nondestructive, and noninvasively fine tuned to each patient's individual needs. Recently, many international groups have begun to apply DBS for the treatment of more complex disorders of the brain, including the neuropsychiatric disorders obsessive–compulsive disorder (OCD) (Goodman et al., 2010) and Tourette syndrome (TS) (Visser-Vandewalle et al., 2004, 2006; Visser-Vandewalle, 2007; Servello et al., 2008). Prior to the option of using neuroprosthetics as viable therapeutics, many neuropsychiatric disorders were managed by behavioral or pharmacological intervention. Behavioral approaches use cognitive behavioral therapy or habit reversal training and are the first strategy when symptoms such as those encountered in OCD become burdensome. Pharmacotherapy for moderate to severe neuropsychiatric disorders are handled on a patient-by-patient basis and typically uses antipsychotics, stimulants, adrenergic inhibitors, or serotonin and norepinephrine reuptake inhibitors. These methods of pharmacotherapy are not specific because they affect many systems of the brain and body and often lead to unwanted side effects. Many rounds of trial and error are needed to find the right combination of medications that are at the right dosages to achieve the desired effect. In situations where the medication level is high to control symptoms, the side effects such as motor complications, cognitive problems, nausea, and hypotension can outweigh the benefits. In cases of severe and medication refractory neuropsychiatric disorders, ablative brain lesions have been effective (Hassler and Dieckmann, 1970; Temel and Visser-Vandewalle, 2004). However, concern about the invasiveness of these procedures and the nonreversible outcome has sparked the need for alternative therapies.

10.2 DBS as a foundational technology

DBS is a foundational technology that has its roots in cardiac pacing. Like a cardiac pacemaker, DBS uses a surgically implanted medical device that has a stimulator circuit, onboard battery, and controller mechanism to deliver carefully patterned electrical stimulation to precisely targeted areas in the brain. Traditional DBS consists of a lead or electrode, the lead extension, which connects the electrode to the device, and the neurostimulator (implanted pulse generator). DBS therapy for the brain leveraged knowledge and regulatory approval strategies from cardiac pacemakers and emerged as an alternative for surgical resection, which prior to DBS development was one of the primary methods for treating a variety of neurological disorders. The mechanism of DBS is unknown; however, electrical stimulation is used to override abnormal neuronal activity within specific brain regions to bring controlling circuits into a more normal state of function, thereby reducing the disorder's symptoms. DBS was originally developed as a replacement for pallidotomy and was presumed to be inhibitory in nature (Vitek, 2002). However, electrical stimulation when programmed properly can produce activation or inhibition of neuronal tissue. This apparent dichotomy has led to a variety of theories for mechanisms, not all of which are exclusory (Vitek, 2002; Wong et al., 2008; Mann et al., 2009). The most common investigations into DBS mechanisms typically evaluate the effect of stimulation on local spiking behaviors. These studies reveal that DBS is likely preferentially activating fiber bundles more so than cell bodies (Rosa et al., 2012). However, the implications for this activation on a network level are still speculative and depend on the electrode type, size, and relative neuro-anatomy (Molnar et al., 2005). Other theories posit that an "informational lesion" exists after stimulating because the pattern of the activation of repetitive computer controlled stimulation is quite different from spontaneous spiking activity typically seen in electrophysiological studies (Walker et al., 2012). Although naturally occurring spike trains have an interspike interval that approximates a Poisson distribution, artificially induced spike activity is extremely regular. Lastly, the resonant effect hypothesis states that DBS may be working by driving local networks at specific resonant frequencies that are inherent properties of existing neuronal networks, and that driving the network at either resonant or dissonant frequencies may have functionally disparate results.

10.2.1 Fundamentals of effective DBS— first-generation approaches

The efficacy of DBS has been demonstrated in many medication refractory hyperkinetic movement disorders by electrically stimulating deep brain structures such as the subthalamic nucleus (STN), the internal segment of the globus pallidus (GPi), or the ventral intermediate nucleus of the thalamus (Vitek, 2002; Montgomery and Gale, 2008; Mann et al., 2009; Zhuang et al., 2009; Lee et al., 2011). DBS was initially FDA approved for ET in 1997, PD in 2002, and dystonia in 2002. The therapy has been used safely for more than 20 years and in more than 100,000 patients worldwide. With proper patient selection and device programming, DBS has enabled improvement in standard scales/measures of disease, quality of life measures, comorbid conditions, medication intake, and chronic care costs. This success has led to the adaptation of DBS as an experimental new therapy for other indications, including neuropsychiatric disorders.

10.2.1.1 Lead localization Effective DBS treatment requires several factors. The most important factor in patient outcomes is lead localization (Montgomery and Gale, 2008). The first level of lead localization involves target selection and surgical placement of the lead within the selected target region (Houeto et al., 2005; Jankovic and Kurlan, 2011). Small variation in lead location can result in changes to the field of activation. If a field of activation includes a nearby nucleus or fiber tract, the stimulation of the DBS lead may result in the activation of these regions and consequently unwanted side effects. The more common side effects for DBS include eye pulling, slurred speech, and trouble with gait. Although a relative consensus exists for optimal targeting for many movement disorders like PD and ET, this is not the case for novel application of DBS for neuropsychiatric disease (Temel and Visser-Vandewalle, 2004). DBS for neuropsychiatric disorders is still experimental, and multiple targets are currently being investigated. New candidate targets can include centromedian (CM) thalamus and nucleus accumbens.

10.2.1.2 Trial-and-error programming The second most important factor for achieving positive outcomes with DBS therapy is programming of the device to efficacious parameters (Porta et al., 2009; Hariz and Robertson, 2010). Unlike anatomical neurosurgery, which accomplishes its goals through gross modifications to nervous system structure, functional neurosurgery works by influencing the pattern of the activation of neuronal tissue (Vitek, 2002; Piedad et al., 2012). The precise response of an individual patient's neuronal tissue to a given electrical stimulation can vary greatly. This variability necessitates the use of rigorous programming paradigms to empirically determine the optimal stimulation parameters to achieve maximum clinical benefit with minimum side effects. The approach to this involves the selection of the appropriate electrode to allow for the adjustment of the site of stimulation. The adjustment of the stimulation parameters allows control over the amount of therapy. Figure 10.1 provides an overview of programming for conventional clinical DBS. Electrodes in current medical practice consist of four contacts, which are numbered from zero to three starting with the most distal tip. Depending on the functionality of the pulse generator, stimulation can occur in monopolar or bipolar mode. In monopolar mode, only one electrode provides stimulation, and the return path for this stimulation is usually defined as the metal packaging or "can" of the neurostimulating device. By contrast, bipolar mode has a local pair of stimulating electrodes where the return path can be adjacent to any of the other three remaining electrodes. Because bipolar stimulation is local, it allows one to limit the spread of activation and is beneficial for shaping the field of stimulation. The main objective of DBS programming is to deliver therapy to the brain target of interest by avoiding the stimulation of the surrounding structures if they are not involved in the disorder. Traditional trial-and-error DBS programming is very inefficient and can take as much as 6 months to obtain the best settings. Many patients also require

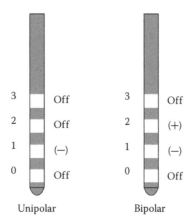

Figure 10.1 Conventional DBS lead selection and configuration for stimulation.

concurrent medication adjustments because they can also affect the stimulation effects. Trial-and-error DBS often begins with an identification of the patient's prominent symptoms and a review of the expectations of DBS. Starting with a monopolar setting, the clinician can attempt to identify the anatomy and position of the electrode within the target nuclei by slowly increasing the threshold of stimulation and observing the apparent change in symptoms. Once gross evaluation is made and side effects are identified, the stimulation can be refined to bipolar if necessary. It is common strategy to make one parameter change with little adjustments to other parameters. For example, one can adjust the amplitude with little or no change to the rate or pulse width or adjust the pulse width with little or no change to the amplitude or rate. Moving beyond trial-and-error programming involves more sophisticated engineering that can simultaneously optimize against all of the stimulation parameters and in conjunction with the behavioral effects. These changes in the traditional clinical paradigm will be described in the next sections.

10.2.1.3 Schedule of stimulation The final factor for effective DBS is the schedule of stimulation (Okun et al., 2013). Many disorders for which DBS is applied have symptoms that wax and wane, or are paroxysmal with symptoms coming in bouts. An ideal stimulation paradigm will administer or amplify stimulation when symptoms are present and disable or reduce stimulation when symptoms are absent. Such paradigms are difficult to develop because of their reliance on close monitoring of the subjects symptoms and the dependency of stimulation paradigms on subjective reports of symptom severity.

10.3 Shifts in research/practice paradigms

Despite recent advances, clinicians and researchers remain limited by the tools available to study, understand, and treat systems of the brain. To achieve maximum benefit, clinicians are often forced to complete a slow, repetitive, and imprecise cycle of observing behaviors and fine-tuning drug or behavioral therapy until the effects of a disease are reduced. To this point, the science has been largely based on a century of identifying associations between features of complex behaviors and diffuse understanding of the brain. Investigators have established these links between specific anatomic regions and complex behavior through a century of neuroscience and neuropsychiatry experiments involving imaging, surgery, and observations of alterations in mental and behavioral state after trauma. Although there is no question that anatomy and behavior are functionally linked, there is a growing body of evidence to suggest that many neural and behavioral processes are not anatomically localized but are emergent from systems that span regions of the brain. Therefore, it may be possible to take an investigative approach that establishes the characteristics of these distributed systems and attempts to restore total brain function. It is possible that total brain function may be restored through multiregion recording and stimulation, appropriately observing neural dysfunction, modeling the relationship between neural systems and behavior, and applying an intervention through neural actuation.

New research and clinical paradigms seek to move beyond this limited understanding to create interventions based on insights that can be gained from the intersection of neuroscience, neurotechnology, and clinical therapy. The goal is to alleviate the burden of psychiatric and neurological disease through multiregion recording and stimulation by observing neural dysfunction, by modeling the relationship between neural systems and behavior, and by applying an intervention through neural actuation. These new efforts have the potential to shift DBS research and clinical practice toward addressing the idea that therapeutic stimulation may not need to be delivered as a continuous series of pulses and could potentially be responsive (Morrell, 2006; Sun et al., 2008; Morrell and RNS System in Epilepsy Study Group, 2011).

10.3.1 Prototype methods and interventions— second-generation DBS

Two prototype devices have emerged to address the changing needs of interfacing with the brain across multiple networks and with evolving stimulation patters. They include the Activa PC+S from Medtronic and the NeuroPace Responsive Neurostimulator. In both systems, the neurostimulator has been specifically designed to collect cortical potentials from cortical sites (electrocorticogram [ECoG] and local field potential [LFP]) and from deep brain electrodes. They can deliver a scheduled paradigm (on for a defined number of seconds and off for a defined number of seconds), a continuous (on all the time) stimulation paradigm, or a responsive paradigm (responds to a physiological signal that is detected by the device). Scheduled and responsive brain stimulations have potential advantages compared with continuous DBS (Okun et al., 2008). These advantages include (1) the provision of a better and specifically tailored approach for individual patients, (2) the ability to address the paroxysmal nature of the symptoms, and (3) a long-term strategy that may prevent or limit tolerance to the stimulation.

Both neurostimulator systems are versatile because they have the ability to acquire, process, and store neural activity from multiple leads in a user defined configuration. The systems include the neurostimulator itself, the implantable leads (up to 4), a programmer that includes a wand and a telemetry interface, and a patient charger. The programmer is used to set up the device, including stimulation setup and recording, as well as to retrieve data for subsequent review. Current electrode interfaces consist of a flexible, isodiametric lead body that encloses four insulated wires and have four cylindrical electrodes at the distal end. The proximal end has four contacts adapted for connecting to the neurostimulator device (i.e., Medtronic lead models 3387). The model 3387 DBS lead has four electrodes with 1.5-mm contacts and 1.5-mm spacing. Another commercially available lead is the Resume II (Medtronic model 3587A) ECoG strip lead, which has a flexible, isodiametric lead body that encloses four insulated wires and has four disk electrodes within a paddle at its distal end. With these two options, a variety of cortical and subcortical targets can be accessed to provide an interface through which stimulation can be delivered or activity of the brain can be monitored by the device or observed by a clinician using a programmer (Figure 10.2).

10.4 Second-generation experimental paradigms— application of DBS for Tourette syndrome

Tourette syndrome (TS) was first described by Gilles de la Tourette in 1885 (Tourette, 1885). TS is a common neuropsychiatric disorder characterized by repetitive involuntary behaviors known as tics (American Psychiatric Association, 2000; Leckman, 2002; Center for Disease Control and Prevention, 2009; Cavanna and Termine, 2012). Tics occur in both motor and phonic manifestations, with the most culturally salient variety being coprolalia, the exclamation of obscenities. However, coprolalia is quite rare, with more common tics manifesting as simple movements involving a single muscle or complex movements involving multiple muscle groups that may seem purposeful but out of context (Cavanna and Seri, 2013). Examples of common simple motor tics include eye blinking, facial contortions, and shoulder shrugging and are typically quick and purposeless. A common phonic tic may consist of coughing, throat clearing, or grunting. By contrast, complex motor tics can appear as any of a wide variety of behaviors and are often self injurious, distressing, and painful. Complex motor tics involve more than one muscle and are usually slower than simple tics. These tics may seem like a purposeful action, such as a grooming behavior or copying an observed behavior. Like complex motor tics, complex phonic tics are often slower and more deliberate expressions that may involve meaningful words and brief sentences.

Diagnosis of TS is accomplished solely by behavioral evaluation. A positive diagnosis requires the presence of at least one phonic tic and at least two motor tics occurring consistently but not necessarily at the same time within a year. Furthermore, these tics cannot be caused by a foreign substance or another medical condition. Although complex tics are not uncommon, diagnosis does not require one to be present. TS occurs in all races and ethnicities but is four times more prevalent in males, with total prevalence at approximately 1% of young children (Robertson et al., 2009). Symptomatic onset of disruptive tic behaviors typically occurs in adolescence before the age of 18 years, with an average age of onset between 5 and 7 years. Motor tics usually appear first between the ages of 3 and 8 years, with phonic tics occurring later around 10. However, most cases

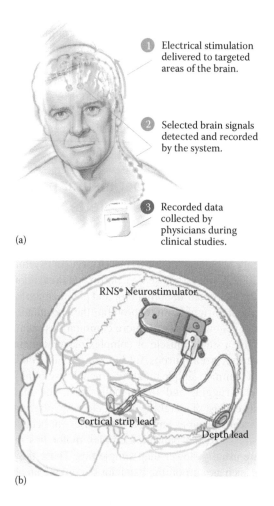

(a)

1 Electrical stimulation delivered to targeted areas of the brain.

2 Selected brain signals detected and recorded by the system.

3 Recorded data collected by physicians during clinical studies.

RNS® Neurostimulator

Cortical strip lead

Depth lead

(b)

Figure 10.2 Second-generation DBS neural interfaces have the capability to sense neural activity, analyze the signals, and provide responsive stimulation. (a) Medtronic PC+S. (From Stanslaski, S., P. Afshar, P. Cong, J. Giftakis, P. Stypulkowski, D. Carlson, D. Linde, D. Ullestad, AI-T. Avestruz, and T. Denison, Design and validation of a fully implantable, chronic, closed-loop neuromodulation device with concurrent sensing and stimulation. *IEEE Trans. Neural Syst. Rehabil. Eng.*, 20, 410–21, 2011; Rouse, A. G., S. R. Stanslaski, P. Cong, R. M. Jensen, P. Afshar, D. Ullestad, R. Gupta, G. F. Molnar, D. W. Moran, and T. J. Denison, A chronic generalized bi-directional brain–machine interface. *J Neural Eng.*, 8, 036018, 2011.) (b) NeuroPace Responsive Neurostimulator. (From Morrell, M., Brain stimulation for epilepsy: can scheduled or responsive neurostimulation stop seizures? *Curr. Opin. Neurol.*, 19, 164–8, 2006.)

(~70%) of TS spontaneously remit by adulthood (Cavanna et al., 2009; Robertson, 2012). Severity of tics can wax and wane throughout the lifetime of the individual and can often be voluntarily suppressed for brief periods. This suppression is usually followed with a period of elevated tic severity. In addition, the vast majority of TS patients report a "premonitory urge" that occurs shortly before a tic occurs (Cavanna and Nani, 2013; Crossley et al., 2014). The premonitory urges consist of the impression of building "tension" that is released by the expression of the tic. This has lead to the term "unvoluntary" to be used to describe the parity between an involuntary urge to action and a voluntary execution of the action.

10.4.1 Comorbidities and pathology

In addition to tic burden, the majority (80–90%) of TS patients also experience symptoms of other neuropsychiatric disorders. The most common of these disorders are OCD and attention deficit/hyperactivity disorder (ADHD), with approximately 60% of TS patients also having one or both comorbidities (Bloch et al., 2006; Scharf et al., 2012; Wanderer et al., 2012). Furthermore, complex tics occasionally include ritualistic behaviors that may lead them to be misdiagnosed as compulsions. These observations have contributed to the view of TS as a "spectrum disorder" that likely shares etiological characteristics with OCD, ADHD, and other tic disorders (Kurlan et al., 2002; Cavanna et al., 2009). Anxiety and affective disorders are among the other common comorbidities of TS patients. Given the profound effect that tics can have on social development, depression and anxiety are likely consequences of living with a disabling condition such as TS. Furthermore, common medications for TS include dopamine antagonists, which themselves can lead to depression.

Although TS is an inheritable disorder, little is known about the underlying pathology of tic expression. The pattern of inheritance for TS is complex—with incomplete penetrance, variable expressivity, and no definitive candidate genes (McNaught and Mink, 2011; Deng et al., 2012). It is likely that a variety of environmental, genetic, and epigenetic mechanisms play a role in TS etiology.

There is no difference in gross brain anatomy between a TS patient and a healthy control that is diagnostic; however, several studies have demonstrated subtle and disparate changes in brain volume in a variety of areas. In two such studies, TS patients showed slight but significant reduction in caudate volume (Peterson et al., 2003; Ludolph et al., 2006). In addition, longitudinal studies have shown that the magnitude of this reduction in volume predicted tic severity later in life (Bloch et al., 2005). However, several studies have shown no difference in striatal volume (Roessner et al., 2009; Miller et al., 2010). Similarly intriguing but inconsistent results are found for volume studies focused on cortical, thalamic, and limbic regions (Ganos et al., 2013). Changes in brain volume tend be primarily observed across brain structures associated with attention, motivation, and cognitive processing (Peterson et al., 2003; Plessen et al., 2009; Draganski et al., 2010). These changes include decreased volume measurement from operculum, ventrolateral prefrontal cortex, rostral anterior cingulate cortex, and orbitofrontal cortex, with increased volume measurements from thalamus and putamen. However, these studies are rarely longitudinal, making it difficult to determine if these changes are causative or compensatory. This variation in results between studies is probably explained best by a small effect size and differences in subject demographics, scanning protocols, or statistical analyses. Nonetheless, these results provide a starting point for further functional imaging studies. The etiology of this disease therefore remains poorly understood, with genetic, developmental, and environmental components likely contributing to a multifactorial etiology.

10.4.2 Basal ganglia and repetitive behaviors

Although the mechanisms of tic generation are unknown, a preponderance of evidence suggests a network dysfunction within prefrontal and subcortical circuits responsible for motor program selection, as shown in Figure 10.3 (Jankovic and Kurlan, 2011; Ganos et al., 2013). Postmortem neuropathological studies

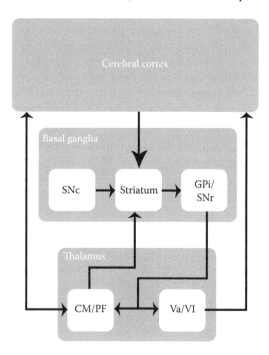

Figure 10.3 Basal ganglia skeletomotor circuitry. A schematic description of the inputs and outputs of the cortico-basal ganglia-thalamocortical loop involved in motor execution.

188

demonstrate a reduction in parvalbumin containing GABAergic cells within the caudate nucleus, putamen, and globus pallidus externus (GPe) of TS patients (Kataoka et al., 2010). A corresponding increase in parvalbumin containing cells within the globus pallidus internus (GPi) was also observed—suggesting a failure of tangential migration from the ganglionic eminence rather than gross neuron death. The hypothesized role of this neuronal population in action selection within the basal ganglia GABAergic system has given rise to the hypothesis that dysfunction in inhibitory microcircuitry may lead to the disinhibition of tic behaviors at the level of the striatum (Mink, 2001). However, a variety of other pathologies are also observed. Investigation of monoamine signaling using positron emission tomography competitive binding assays showed decreased dopaminergic binding in the cingulate and various frontal regions, while increased dopamine release and serotonin binding potential was observed in the striatum (Leckman et al., 2010; Steeves et al., 2010). Dopaminergic activity in this region is involved in both limbic and motor circuits, making dopamine dysfunction a reasonable candidate component of TS pathophysiology. Furthermore, these observations vary by study, with large differences in patient populations. The pathology of TS therefore remains heterogeneous and poorly understood. This pathological heterogeneity is reflected in the spectrum of symptom expression and the variable efficacy of pharmacological treatments.

This observed dysfunction within basal ganglia circuitry has led to current models of TS and other repetitive behavior disorders. Although no animal models of TS capture the phenomenology of the urge to tic, tic-like behaviors can be induced by striatal disinhibition. Injection of bicuculline into the striatum results in the suppression of GABA signaling and ultimately increased activity in downstream motor circuitry. This produces sudden jerking movements that resemble tics. This model directly links dysfunction within the striatum to the appearance of motor tics. An evaluation of electrophysiological recordings throughout the cortico-striato-thalamocortical loop has shown that motor tics induced in this way are associated with phasic changes of neuronal firing patterns throughout the network.

Although animal models of tic can be valuable research assets, they fail to account for the occurrence of complex tics, urge to tic, and stereotyped behaviors. A successful model of TS must account for these unique characteristics. One such model emerges with the observation that the focal activation of a set of striatal neurons can also result in stereotyped motor output. The basal ganglia is hypothesized to play a role in motor program selection, and in this model, multiple motor programs are simultaneously considered until a desired behavior is selected, whereupon the remaining motor programs are suppressed. If the aberrant activation of a subset of striatal neurons causes pathological inhibition of output neurons associated with a desired behavior, competing motor behaviors would have a lower threshold for activation. Consequently, tic behaviors and other nondesired behaviors may be likely to occur in this environment.

10.4.3 CM thalamus

The optimal target for DBS therapy in TS is still under investigation (Temel and Visser-Vandewalle, 2004; Ackermans et al., 2006). Although GPi and nucleus accumbens are also being explored, the CM nucleus of the thalamus is an efficacious target in many TS patients and is the target of choice for our surgical center. This region is a nonspecific thalamic nucleus that provides behaviorally relevant sensory information to basal ganglia circuitry (Matsumoto et al., 2001; Minamimoto and Kimura, 2002). This information drives motor program selection on the basis of selective attention; the anatomy of this region reflects the diversity needed for the multimodal integration required for attentional orienting.

The CM region has massive reciprocal connections with the striatum and widespread projections to GPi, amygdala, nucleus accumbens, and sensorimotor cortices (Raeva, 2006; Vertes et al., 2012). Primary inputs come from the superior colliculus, GPi, locus coeruleus, and thalamic reticular nucleus (Berendse and Groenewegen, 1991). These connections are divided along the rostral/caudal axis of the nucleus, with the rostral portion primarily connected to limbic circuitry and the caudal portion connected with sensorimotor areas (Minamimoto et al., 2009). This positioning allows the CM nucleus to uniquely affect the action-gating circuits likely involved in TS symptomatology by affecting both motor and limbic components of action selection within the basal ganglia.

On the basis of this connectivity and the efficacy of CM DBS in TS, there is a high probability that neuronal signals in this region will carry information relevant to tic genesis (Marceglia et al., 2010). The presence of a DBS device allows the quantification of these signals by recording the activity of extracellular action potentials intraoperatively and LFPs postoperatively with appropriate technology. This information is an important and rarely explored venue of analysis for elucidating the pathophysiology of tic generation, which is otherwise elusive. Investigation of this region using LFP analysis has the opportunity to not only describe tic-related physiology but also provide an avenue of investigation into how this brain circuitry contributes to motor behaviors in healthy humans.

10.4.4 New experimental test beds

Understanding the physiological underpinnings of motor and vocal tics is a complex and crucial step in understanding the pathophysiology of TS. It is essential to combine high spatial and temporal resolution physiological recordings of TS subjects postoperatively. This allows for answering fundamental questions about brain activity surrounding tic formation in both target regions, as well as the opportunity to examine system interaction in this disease state during symptom expression. In addition to the standard clinical assessment, new experimental test beds can be used to study the relationship between LFPs and tic occurrence.

A summary of an example system (positional sensor, video recording, EEG, and DBS system) is shown in Figure 10.4.

Brain electrophysiology from TS patients in the postoperative state can be sampled using the following paradigm. LFP recordings are compared across sessions where the subject is either (1) experiencing spontaneous tics or (2) voluntarily imitating their tic movements. Behavior during tic, non-tic, and imitation-tic actions is recorded in synchrony with the LFPs. Non-tic and imitation tic actions will be chosen to provide baseline physiological and kinematic controls for true tic comparison. Synchronized video recordings are used to determine the onset of vocal tics and motor tics occurring in the body. The synchronization of electrophysiology with measurements of behavior can then be used to correlate neuronal modulations with behavior (i.e., tic). In the example system shown in Figure 10.4, thalamic LFPs were transmitted wirelessly. The implanted processor buffered recorded data, which were retrieved via a telemetry wand connected to a computer. Positional data are collected via an Ascension Trakstar system (Burlington, VT) connected to a Tucker-Davis Recording system (TDT, Alachua, FL). In the examples shown here, the Ascension Trakstar sensor is attached to the patient's finger and records the position of the arm end point represented as x, y, and z coordinates in three-dimensional space (sampling rate 400.2 Hz). To time synchronize the positional information with the neuronal recordings, a single electrode is placed directly over the implanted pulse generator and connected to the TDT recording system. A brief stimulation pulse (duration 100 ms) is used to time synchronize the data from the DBS device to the measures of behavior. Positional data, EEG, and video recordings are synchronized and stored on a second dedicated computer.

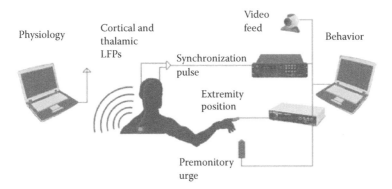

Figure 10.4 New DBS experimental setups are composed of two elements. LFPs are recorded on the implanted DBS device and transmitted wirelessly to an offline computer for further analysis. Behavioral assessment is accomplished using live video feed and limb position. This information is synchronized with physiological data.

191

10.4.5 Physiological biomarkers of TS

The success of surgical treatments in the CM thalamus supports the notion that this region likely contributes to a network dysfunction responsible for tic generation (Xiao et al., 2009). The CM is an intralaminar thalamic nucleus that has reciprocal connections with the action-gating pathways of the basal ganglia. This region has been demonstrated to process relevant sensory, executive, and arousal information that is important for motor planning (Matsumoto et al., 2001; Raeva, 2006). The electrical stimulation of fiber bundles within this region is likely to produce action potentials. Downstream effects of chronic electrical stimulation are likely to manifest throughout the motor circuitry and can result in the functional reorganization of circuit architecture through Hebbian-based mechanisms (Steriade et al., 1996; Schreckenberger et al., 2004). These functional changes may manifest on both short- and long-term time scales, depending on whether the acute effects of stimulation are more important than the lasting functional changes that may be occurring. If functional reorganization is occurring, it will produce changes in resting state and functional activity that can be measured by analyzing the LFP (Steriade et al., 1991; Bront-Stewart, 2011).

Neuronal population activity producing tic phenomenology is likely encoded by network oscillations and their dynamics (Fries, 2005). For example, the presence of gamma band oscillations reflects a state of high neuronal excitability and synchronization by cell ensembles, which are critical for neuronal communication and normal brain function (Brovelli et al., 2005; Hughes, 2008; Fries, 2009; Buzsáki et al., 2012). Abnormal gamma oscillations have been implicated in such disorders as PD, schizophrenia, and depression (Llinas et al., 1999; Uhlhaas and Singer, 2006, 2010). However, very little is known about the functional significance of these oscillations in TS, aberrant or otherwise, or how they are modulated by DBS therapy (Leckman et al., 2006).

The goal of the experiments shown here is to describe the electrophysiology and clinical benefits of DBS in five human TS subjects implanted with NeuroPace neurostimulators (NeuroPace, Mountain View, CA). This novel experimental and clinical paradigm enabled the capture of LFPs from the CM thalamic region and their relationship to tic expression over the course of a 6-month clinical trial, thus providing a long-term study of thalamic neurophysiology in human TS subjects undergoing DBS. Electrophysiology is coupled with tic phenomenology to correlate chronic, temporal changes of neural activity with clinical evaluations of tic expression.

10.4.6 Surgery and subjects

Five of nine subjects shown here met the *Diagnostic and Statistical Manual of Mental Disorders, 4th Edition, Text Revision* criteria for TS as well as the criteria

192

for TS DBS proposed by the Tourette Syndrome Association and were enrolled in this study ("Definitions and classification of tic disorders," 1993; American Psychiatric Association, 2000) (FDA clinical trial NCT01329198, clinicaltrials .gov). Three females and two males between 28 and 39 years (mean, 34.4 years) all had severe motor and phonic tics with disease duration at the time of implantation of 20 to 37 years (mean, 28.8 years), as shown in Table 10.1. Tic severity was assessed at screening, before surgery, and at each subsequent visit using the Yale Global Tic Severity Scale (YGTSS) (Leckman et al., 1989) and the Modified Rush Tic Rating Scale (MRTRS) (Goetz et al., 1999). All subjects had comorbidities with OCD symptoms. All subjects were refractory to traditional medications and were approved as candidates for DBS surgery by an interdisciplinary expert panel and an independent ethics panel.

To implant DBS electrodes into the CM thalamus, stereotaxic neurosurgery was performed. Localization of the CM thalamus was achieved by the use of a 3-T MRI scan, an FGATIR MRI protocol, and a morphable superimposed atlas (Sudhyadhom et al., 2009, 2012). Intraoperative microelectrode recordings were

Table 10.1 Subject Demographics

Implant Number	Sex	Age	Disease Duration	Common Tics	Behavioral Comorbidities
TS1	F	34	26	Head jerks, limb-jerking, slapping/hitting self and hitting nearby objects, abdominal tensing, coprolalia	OCD moderate and chronic
TS2	M	37	34	Eye rolling, rotating wrists and shoulders, cracking joints, hitting nearby objects, vomiting	ADHD hyperactive and impulsive, stable, and secondary substance dependency, OCD traits
TS3	M	28	20	Face scrunching, arm jerks, head twists, bending at the waist, copropraxia, squawking, grunting, sniffing	OCD, moderate and chronic
TS4	F	39	37	Eye rolling, jaw cracking, head twists, fingertip tapping, hits with elbow, copropraxia, growling, coprolalia	OCD mild to moderate and chronic, PTSD mild and chronic (resolved at time of DBS)
TS5	F	36	27	Fingertip waving, grimacing, eye rolling, echolalia, yelling, and growling	OCD current moderate and chronic, PTSD (past, resolved), MDD (past, resolved)

Note: Components from this table have been reproduced with permission from Okun et al. (2013) and have been published with the original NIH-supported FDA clinical trial (clinicaltrials.gov).

performed to assess and confirm the target location by identifying characteristic neuronal firing patterns along the electrode trajectory as well as "border cells" at the target. The microelectrodes (FHC Inc., Bowdoin, ME) were advanced to the target using a micropositioner (FHC Inc.), and electrophysiological recordings were collected in real time. Once the optimal implantation area was established, the DBS macroelectrode was inserted, and the NeuroPace neurostimulator was connected and fixed in the cranium. Test LFPs were recorded in the operating room after the internalization of the leads to evaluate the functionality of the implanted hardware.

Three of the five subjects enrolled in the study received two unilateral pulse generators each controlling ipsilateral leads. One subject (TS1) initially received a single generator controlling bilateral leads but had two replacement generators implanted after 4 months, each serving unilateral leads. One subject (TS2) received a single generator serving bilateral leads for the duration of the study. LFPs were recorded from macroelectrodes with cylindrical contacts (1.27 mm diameter, 2.1 mm length, 3.5 mm center-to-center spacing, and 500–600 Ω average impedance) implanted in the CM region.

Subjects were randomized for initial postoperative stimulation onset. Thus, TS1 and TS3 received stimulation beginning with the first visit (i.e., 1 month postoperatively), and TS2, TS4, and TS5 started stimulation at the second visit (i.e., 2 months postoperatively). Stimulation was delivered by alternating brief periods of on and off stimulation conditions (scheduled stimulation paradigm). Settings, including lead polarity, current, pulse width, frequency, and duty cycle, were customized at each visit for each subject. Stimulation settings among the subjects had ranges from 1 to 4.50 mA current, from 80 to 240 μs pulse width, from 100 to 149.2 Hz frequency, and from 17% to 50% duty cycle. All subjects were stimulated with a unipolar setting on contact 1 (the most ventral/deepest contact) with the exception of TS2, who was stimulated on contact 2. This decision was based on optimal clinical responses.

10.4.7 Lead localization

A CT scan was performed 1 month postoperatively and fused to the preoperative MRI to confirm lead placement, as shown in Figure 10.5. Lead contacts were localized to the CM region of the thalamus with the most distal contact located approximately 3 mm ventral to the target region. Lead spacing was 3.5 mm (center to center), resulting in contacts 1, 2, and 3 lying within the general target region and contact 4 lying dorsal. The leads for subject TS1 were found to lie slightly more anterior than the rest of the cohort; the leads for subject TS5 were found to lie more ventral. The position in the x coordinate is measured medial/lateral relative to midline, y is anterior/posterior relative to midcommissural point, and z is dorsal/ventral relative to the commissural plane.

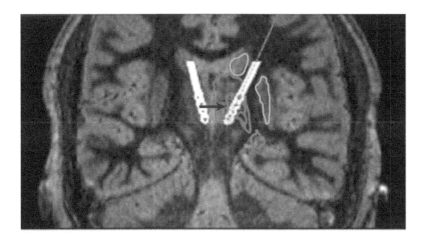

Figure 10.5 Lead location using CT-MRI fusion. Black arrow represents the target location (CM thalamus). Thalamus is outlined in dark-gray, striatum is outlined in light-gray, subthalamic nucleus is outlined in black, and lead trajectory is represented by the dashed gray line.

10.4.8 Data collection

All LFP recordings were performed while the stimulator was in the off state. Follow-up assessments consisted of monthly visits to the clinic for the first 6 months after implant surgery. Subjects TS2, TS3, and TS5 were recorded approximately monthly for 6 months postoperatively. Subject TS1 was recorded on or around postoperative months 1, 2, 3, and 4. After 4 months of stimulation, this subject underwent a second surgery to remove the first device and to implant two new devices, each serving the original ipsilateral leads. She was reenrolled and resumed testing protocols on postoperative months 13 and 14. For her initial 6-month follow-up, subject TS4 was recorded postoperatively at months 1, 2, 3, and 5 due to travel difficulties. Long-term follow-ups every 6 to 12 months were planned for all subjects.

Four channels of thalamic LFPs from each device were filtered (3–125 Hz, 12 dB/oct) and digitized at 250 Hz from two contact pairs. For pulse generators serving unilateral leads, as shown in Figure 10.6a, channel 1 represents LFP recordings across contacts 1 and 2, channel 2 represents contacts 2 and 3, channel 3 represents contacts 3 and 4, and channel 4 represents contacts 1 and 4. For pulse generators serving bilateral leads, as shown in Figure 10.6b, channels 1 and 2 represent the LFP recordings across contacts 1 and 2 and contacts 3 and 4, respectively, whereas channels 3 and 4 represent the LFP recordings from contacts 1 and 2 and contacts 3 and 4, respectively, in the contralateral hemisphere.

Recordings were made during several conditions. For the baseline condition, subjects were recorded with the stimulator off and while resting comfortably and

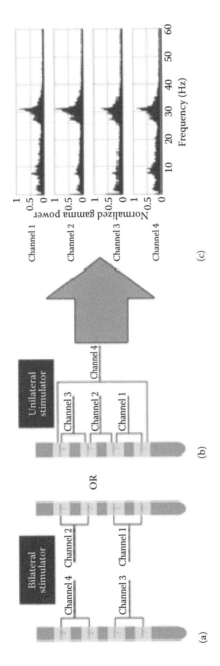

Figure 10.6 Channel diagram for LFP collection. (a) Schematic for channels representing a subject who has a single stimulator controlling bilateral leads. (b) Schematic for channels representing a subject who has two stimulators each controlling ipsilateral leads. (c) Representative power spectra from a bilaterally implanted patient showing changes in spectral density as a function of channel number.

behaving naturally. They were instructed to not suppress the urge to tic so that tic expression during baseline was unimpeded. As such, baseline consisted of periods, which were tic-free in combination with periods of tic expression. These recordings were taken at each visit and typically lasted for a period of 2 minutes and did not exceed 4 minutes.

Spectral power was estimated using Welch's method (1024 point hamming window, 50% overlap, and 4096 point Fourier transform). For the purposes of quantitative analysis, frequency cutoffs for theta and gamma were defined as 4–7 and 25–45 Hz, respectively. Gamma frequency was determined as optimal power fluctuations within the normal gamma range. Power values were normalized across visits for each subject by setting the sum of all power spectra to unity. A scaling factor of 500 was then applied to all normalized spectra such that mean gamma values were set between zero and one. Mean power was computed for all frequency bands as the average power between cutoff frequencies. Because normalization could cause frequency bias if one frequency band dominates the spectra, comparisons between theta and gamma used peak power values calculated from the prenormalized data. For the analysis of chronic DBS effects on thalamic activity, spectra were calculated for baseline recordings for each subject for each monthly visit. The power of the frequency bands in the spectra of the chronic recordings were then correlated with clinical metrics, which assessed changes in motor tic severity, phonic tic severity, and overall impairment in the prior month.

10.4.9 Clinical assessment

The YGTSS was used to clinically assess subjects on the first day of each of their monthly follow-up visits, and the outcomes recorded represented steady states of motor and vocal tic manifestations. The YGTSS represents the tic activity as reported over several weeks. In addition to the overall score, all subscores of the YGTSS for severity and impairment were evaluated with respect to neurophysiological changes.

The MRTRS was used as the clinical metric for more acute responses to changing stimulation parameters. It was typically evaluated twice daily on each study visit (before and after stimulation parameters were adjusted) to provide continuing information about meaningful changes in the clinical state. In some cases, only one scale per day was available. Measured MRTRS scores were in good agreement with YGTSS scores (Spearman's correlation coefficient = 0.68) and consistent with the literature (Goetz et al., 1999).

10.4.10 Characterization of power spectral density

Spectral analysis revealed a continuum of multiple oscillatory rhythms observed across a broad range of frequencies. Decreasing power was observed with increasing frequency as predicted by power law of natural dynamical systems (DeCoteau et al., 2007). An initial comparison of the effect of pre- and post-DBS on the dynamics of these frequencies indicated fluctuations in gamma band power that were not readily observed in other frequency ranges, as shown in Figure 10.7. These observations motivated further analysis on gamma band activity. Changes in gamma power were observed predominantly between 30 and 40 Hz, with a range varying between 25 and 45 Hz.

10.4.11 Correlation of gamma with clinical improvement

Notable increases in normalized gamma band power were indicative of clinical benefit ($N = 5/5$). Correlation analysis revealed that the power of the gamma oscillations was inversely related to the overall severity of TS symptoms as measured by the YGTSS, as shown in Figure 10.8. Improvement in YGTSS was observed and compared with preoperative screening scores. Over the 6-month study, all subjects showed some clinical benefit from DBS with most subjects showing a strong reciprocal increase in gamma band presence. Subjects TS2, TS3, and TS4 showed the largest negative correlations. TS1 and TS5 showed low correlations and small clinical benefits. TS3 and TS4 were the best responders to therapy (33% and 32% decrease in YGTSS) with substantially greater clinical benefit than TS1 and TS5 (1% and 18% decrease in YGTSS). The cohort showed a positive trend in gamma power negatively correlating with decreasing YGTSS. TS2 generally exhibited minimal responses to DBS and also marginal changes in gamma between months 1 and 6. For all but month 5, this subject's YGTSS scores were consistently high, indicating a low average clinical benefit (18%). In month 5, however, there was a large (41%) improvement in the YGTSS corresponding to a twofold increase in gamma in the right hemisphere. Both gamma power and YGTSS returned to previous levels the following month. TS1 and TS5 showed only a small clinical improvement corresponding to very small changes in gamma power (<10%).

The quantification of theta band activity did not reflect a similar correlation with symptomatology. Although fluctuations in gamma and theta power were observed for all, only the subjects that had large increases in gamma were correlated with large reductions in tic expression. Subjects TS3 and TS4 showed a 670% and 72% increase in gamma, respectively, and also exhibited the best clinical outcomes. By contrast, subjects TS1 and TS5 had large increases in theta (160% and 80%, respectively) with small changes in gamma and derived a correspondingly small clinical benefit. Across months 1 to 5, TS2 exhibited opposing changes in theta band power (−20% and 48%) and yet showed considerable increase in gamma band power (869% and 124%) between hemispheres. Although the greatest change in YGTSS (41%) occurred in month 5, the analysis of activity in the following month revealed a precipitous drop in both YGTSS and gamma. Interestingly, in addition to prominent increases in gamma, the two best responders also showed reciprocal decreases in theta power.

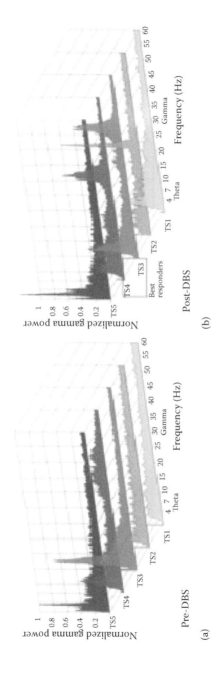

Figure 10.7 Power spectra before and after DBS. (a) Average power spectral density plots from five subjects 1 month after DBS. (b) Average power spectral density plots from five subjects after prolonged (5–6 months) DBS therapy. Pronounced changes in spectral density are observed in the gamma range, which are particularly prominent for those patients who responded best to therapy.

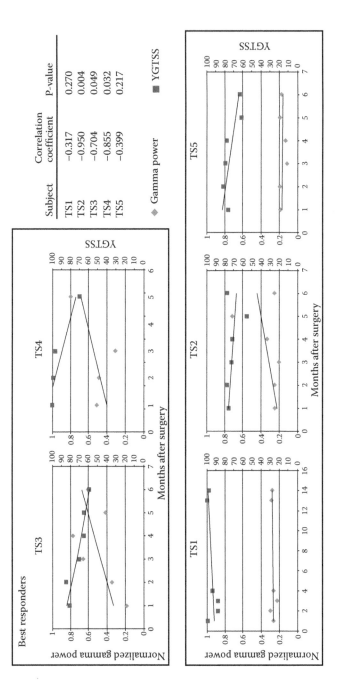

Figure 10.8 Correlation of clinical benefit with gamma power. Plots of gamma power against tic severity (YGTSS) showing strong negative correlation for 3/5 patients.

10.5 Conclusion

The use of direct brain stimulation as a therapeutic intervention for complex disorders of the brain is continuing to evolve. From the first-generation DBS, which was based on simple cardiac pacing, to the second-generation advanced techniques that have the ability to sense, analyze, and stimulate, the novel experimental frameworks and results presented here provide better insight into the development of new therapeutics. At its core, new approaches enable direct translation of neurophysiological modulations to clinical outcomes. It is possible that the identification of additional neural biomarkers that are correlated with brain disorder expression may be used to develop next-generation responsive stimulation paradigms that involve intermittent and time resolved neuromodulation that is directly related to behavior. Although such closed-loop neuroprosthetic therapies must be tailored to individual behaviors, the results from the TS application shown here provide illustrations of the first critical steps in understanding the spectral features of LFPs that may encode the underlying problem and can be used to inform an intelligent response.

A key factor to attaining therapeutic benefit still hinges on target selection, and optimal implantation locations are still under debate. Moreover, there can be a spectrum of clinical benefits that can be due in part to variation in final lead location. To control for this type of variability, methodologies are needed evaluate an individual target region's response to acute stimulation such that clinicians can use this evaluation as a marker for putative surgical targeting sites. Although the results shown here demonstrate symptom relief correlated with increased synchronization within a single oscillatory frequency, multiple frequencies or features may be involved. Modeling is needed to understand the representation and integration of highly circumscribed networks. The modeling knowledge should be applied to advanced control methodologies to elucidate or predict responses to stimulation patterns to guide optimal electrode placement and network effects.

Exercise

1. Compare and contrast the first-generation and the second-generation DBS devices. What are the fundamental differences in the approaches? What are the strengths and weaknesses? What problems are easily overcome with each approach? What are the engineering and neurophysiological challenges of each?

References

Chapter 1

Hubel, D. H. and T. N. Wiesel. 1959. "Receptive fields of single neurones in the cat's striate cortex." *The Journal of Physiology* 148:574–91.

Kelley, T. 2007. *The Art of Innovation.* New York: Crown Business.

Miranda, R. A., W. D. Casebeer, A. M. Hein, J. W. Judy, E. P. Krotkov, T. L. Laabs, J. E. Manzo et al. 2015. "DARPA-funded efforts in the development of novel brain–computer interface technologies." *Journal of Neuroscience Methods* 244:52–67. doi:10.1016/j.jneumeth.2014.07.019.

Shen, H. 2013. "Neurotechnology: BRAIN storm." *Nature* 503 (7474):26–8. doi:10.1038/503026a.

Witkowski, J. A. 1992. "Ramon y Cajal: Observer and interpreter." *Trends in Neurosciences* 15 (12):484.

Chapter 2

Buzsáki, G. 2004. "Large-scale recording of neuronal ensembles." *Nature Neuroscience* 75 (5):446–51.

Collinger, J. L., B. Wodlinger, J. E. Downey, W. Wang, E. C. Tyler-Kabara, D. J. Weber, A. J. C. McMorland, M. Velliste, M. L. Boninger, and A. B. Schwartz. 2013. "High-performance neuroprosthetic control by an individual with tetraplegia." *Lancet* 381 (9866):557–64. doi:10.1016/S0140-6736(12)61816-9.

Freeman, W. J. 2000. "Mesoscopic neurodynamics: From neuron to brain." *Journal of Physiology-Paris* 94 (5–6):303–22.

Hassler, C., T. Boretius, and T. Stieglitz. 2010. "Polymers for neural implants." *Journal of Polymer Science Part B: Polymer Physics* 49 (1):18–33. doi:10.1002/polb.22169.

Hochberg, L. R., M. D. Serruya, G. M. Friehs, J. A. Mukand, M. Saleh, A. H. Caplan, A. Branner, D. Chen, R. D. Penn, and J. P. Donoghue. 2006. "Neuronal ensemble control of prosthetic devices by a human with tetraplegia." *Nature* 442 (7099):164–71.

Hubel, D. H. 1957. "Tungsten microelectrode for recording from single units." *Science (New York, NY)* 125 (3247):549–50.

Jasper, H. and W. Penfield. 1954. *Epilepsy and the Functional Anatomy of the Human Brain.* Boston: Little, Brown and Co.

References

Kandel, E. R., J. H. Schwartz, and T. M. Jessell. 2000. *Principles of Neural Science*. New York: McGraw-Hill.

Kim, D., J. Viventi, J. J. Amsden, J. Xiao, L. Vigeland, Y. Kim, J. A. Blanco et al. 2010. "Dissolvable films of silk fibroin for ultrathin conformal bio-integrated electronics." *Nature Materials* 9 (5):1–7. doi:papers://7964EF6A-BE0A-4F2B-9048-9B5C50D5342A/Paper/p11864.

Kipke, D. R., W. Shain, G. Buzsáki, E. Fetz, J. M. Henderson, J. F. Hetke, and G. Schalk. 2008. "Advanced neurotechnologies for chronic neural interfaces: New horizons and clinical opportunities." *The Journal of Neuroscience: The Official Journal of the Society for Neuroscience* 28 (46):11830–8. doi:10.1523/JNEUROSCI.3879-08.2008.

Kipke, D. R., R. J. Vetter, and J. C. Williams. 2003. "Silicon-substrate intracortical microelectrode arrays for long-term recording of neuronal spike activity in cerebral cortex." *IEEE Transactions on Rehabilitation Engineering* 11 (2):151–5.

Liker, M. A., D. S. Won, V. Y. Rao, and S. E. Hua. 2008. "Deep brain stimulation: An evolving technology." In *Proceedings of the IEEE*, 1129–41. IEEE.

Maynard, E. M., C. T. Nordhausen, and R. A. Normann. 1997. "The Utah Intracortical Electrode Array: A recording structure for potential brain–computer interfaces." *Electroencephalography and Clinical Neurophysiology* 102 (3):228–39.

Nunez, P. L. 1981. *Electric Fields of the Brain: The Neurophysics of EEG*. New York: Oxford University Press.

Raffa, V., V. Pensabene, A. Menciassi, and P. Dario. 2007. "Design criteria of neuron/electrode interface. The focused ion beam technology as an analytical method to investigate the effect of electrode surface morphology on neurocompatibility." *Biomedical Microdevices* 9 (3):371–83. doi:10.1007/S10544-006-9042-2.

Rousche, P. J. and R. A. Normann. 1992. "A method for pneumatically inserting an array of penetrating electrodes into cortical tissue." *Annals of Biomedical Engineering* 20: 413–22.

Salcman, M. and M. J. Bak. 1977. "Design, fabrication, and in vivo behavior of chronic recording intracortical microelectrodes." *Biomedical Engineering* 24 (2):121–8.

Schwartz, A., X. Cui, D. Weber, and D. Moran. 2006. "Brain-controlled interfaces: Movement restoration with neural prosthetics." *Neuron* 52 (1):205–20. doi:10.1016/j.neuron.2006.09.019.

Sejnowski, T. J. and P. S. Churchland. 1989. "Brain and cognition." In *Foundations of Cognitive Science*, edited by M. I. Posner, 302–56. Cambridge, MA: MIT Press.

Thelin, J., H. Jörntell, E. Psouni, M. Garwicz, J. Schouenborg, N. Danielsen, and C. E. Linsmeier. 2011. "Implant size and fixation mode strongly influence tissue reactions in the CNS." *PLoS One* 6 (1):e16267. doi:10.1371/journal.pone.0016267.

Vetter, R. J., J. C. Williams, J. F. Hetke, E. A. Nunamaker, and D. R. Kipke. 2004. "Chronic neural recording using silicon-substrate microelectrode arrays implanted in cerebral cortex." *IEEE Transactions on Biomedical Engineering* 51 (6):896–904.

Viventi, J., D.-H. Kim, L. Vigeland, E. S. Frechette, J. A. Blanco, Y.-S. Kim, A. E. Avrin et al. 2011. "Flexible, foldable, actively multiplexed, high-density electrode array for mapping brain activity in vivo." *Nature Neuroscience* 14 (12):1599–605. doi:10.1038/nn.2973.

Williams, J. C., R. L. Rennaker, and D. R. Kipke. 1999. "Long-term neural recording characteristics of wire microelectrode arrays implanted in cerebral cortex." *Brain Research Protocols* 4 (3):303.

Witkowski, J. A. 1992. "Ramon y Cajal: Observer and interpreter." *Trends in Neurosciences* 15 (12):484.

Wolpaw, J. and E. W. Wolpaw. 2011. *Brain–Computer Interfaces: Principles and Practice*. New York: Oxford University Press.

Chapter 3

Harrison, R. R. and C. Charles. 2003. "A low-power low-noise CMOS amplifier for neural recording applications." *IEEE Journal of Solid-State Circuits* 38:958–65.

Haykin, S. 1996. *Adaptive Filter Theory*, 3rd ed. Upper Saddle River, NJ: Prentice-Hall International.

Nicolelis, M. A. L., D. Dimitrov, J. M. Carmena, R. Crist, G. Lehew, J. D. Kralik, and S. P. Wise. 2003. "Chronic, multi-site, multi-electrode recordings in macaque monkeys." *Proceedings of the National Academy of Sciences of the United States of America* 100 (19):11041–6.

Oppenheim, A. V., R. W. Schafer, and J. R. Buck. 1999. *Discrete-Time Signal Processing*. New York: Prentice-Hall.

Polikov, V., P. A. Tresco, and W. M. Reichert. 2005. "Response of brain tissue to chronically implanted neural electrodes." *Journal of Neuroscience Methods* 148:1–18.

Scott, S. H. 2006. "Neuroscience: Converting thoughts into action." *Nature* 442 (7099):141.

Tomer, R., L. Ye, B. Hsueh, and K. Deisseroth. 2014. "Advanced CLARITY for rapid and high-resolution imaging of intact tissues." *Nature Protocols* 9 (7):1682–97. doi:10.1038/nprot.2014.123.

Winter, D. A. 2009. *Biomechanics and Motor Control of Human Movement*. Hoboken, NJ: John Wiley & Sons.

Wise, K. D., D. J. Anderson, J. F. Hetke, D. R. Kipke, and K. Najafi. 2004. "Wireless implantable microsystems: High-density electronic interfaces to the nervous system." *Proceedings of the IEEE* 92 (1):76–97.

Zhang, J., F. Laiwalla, J. A. Kim, H. Urabe, R. Van Wagenen, Y. K. Song, B. W. Connors, F. Zhang, K. Deisseroth, and A. V. Nurmikko. 2009. "Integrated device for optical stimulation and spatiotemporal electrical recording of neural activity in light-sensitized brain tissue." *Journal of Neural Engineering* 6:055007. doi:papers://7964EF6A-BE0A-4F2B-9048-9B5C50D5342A/Paper/p11639.

Chapter 4

Burman, K. J., S. M. Palmer, M. Gamberini, M. W. Spitzer, and M. G. P. Rosa. 2008. "Anatomical and physiological definition of the motor cortex of the marmoset monkey." *The Journal of Comparative Neurology* 506 (5):860–76. doi:10.1002/cne.21580.

Buzsáki, G. 2004. "Large-scale recording of neuronal ensembles." *Nature Neuroscience* 75 (5):446–51.

Gage, G. J., D. R. Kipke, and W. Shain. 2012. "Whole animal perfusion fixation for rodents." *Journal of Visualized Experiments* (65):3564. doi:10.3791/3564.

Halpern, C. H., U. Samadani, B. Litt, J. L. Jaggi, and G. H. Baltuch. 2008. "Deep brain stimulation for epilepsy." *Neurotherapeutics* 5 (1):59–67. doi:http://dx.doi.org/10.1016/j.nurt.2007.10.065.

Koikkalainen, J., J. Hirvonen, M. Nyman, J. Lötjönen, J. Hietala, and U. Ruotsalainen. 2007. "Shape variability of the human striatum—Effects of age and gender." *Neuroimage* 34 (1):85–93. doi:10.1016/j.neuroimage.2006.08.039.

House, P. A., J. D. MacDonald, P. A. Tresco, and R. A. Normann. 2006. "Acute microelectrode array implantation into human neocortex: Preliminary technique and histological considerations." *Neurosurgical Focus* 20 (5):E4. doi:papers2://publication/uuid/7AC6BEE7-FFC1-4915-B966-C1E628CE8632.

References

Maynard, E. M., C. T. Nordhausen, and R. A. Normann. 1997. "The Utah Intracortical Electrode Array: A recording structure for potential brain–computer interfaces." *Electroencephalography and Clinical Neurophysiology* 102 (3):228–39.

Mollazadeh, M., V. Aggarwal, A. G. Davidson, A. J. Law, N. V. Thakor, and M. H. Schieber. 2011. "Spatiotemporal variation of multiple neurophysiological signals in the primary motor cortex during dexterous reach-to-grasp movements." *The Journal of Neuroscience: The Official Journal of the Society for Neuroscience* 31 (43):15531–43. doi:papers2://publication/doi/10.1523/JNEUROSCI.2999-11.2011.

Musallam, S., M. J. Bak, P. R. Troyk, and R. A. Andersen. 2007. "A floating metal microelectrode array for chronic implantation." *Journal of Neuroscience Methods* 160 (1):122–7. doi:papers2://publication/doi/10.1016/j.jneumeth.2006.09.005.

Nicolelis, M. A. L. 1999. *Methods for Neural Ensemble Recordings.* Boca Raton, FL: CRC Press.

Nicolelis, M. A. L., D. Dimitrov, J. M. Carmena, R. Crist, G. Lehew, J. D. Kralik, and S. P. Wise. 2003. "Chronic, multi-site, multi-electrode recordings in macaque monkeys." *Proceedings of the National Academy of Sciences of the United States of America* 100 (19):11041–6.

Nordhausen, C. T., P. J. Rousche, and R. A. Normann. 1994. "Optimizing recording capabilities of the Utah-Intracortical-Electrode-Array." *Brain Research* 637 (1–2):27–36.

Palazzi, X. and N. Bordier. 2008. *The Marmoset Brain in Stereotaxic Coordinates.* New York: Springer Verlag.

Pennartz, C. M., H. J. Groenewegen, and F. H. Lopes Da Silva. 1994. "The nucleus accumbens as a complex of functionally distinct neuronal ensembles: An integration of behavioural, electrophysiological and anatomical data." *Progress in Neurobiology* 42:719–61.

Perlmutter, J. S. and J. W. Mink. 2006. "Deep brain stimulation." *Annual Review of Neuroscience* 29:229–57. doi:papers2://publication/uuid/15CF07FE-25ED-46B2-BFE1-9029EEC4195E.

Picard, C., A. Olivier, and G. Bertrand. 1983. "The first human stereotaxic apparatus." *Journal of Neurosurgery* 59 (4):673–6. doi:papers2://publication/uuid/67E83C00-E109-4C41-A8A7-DE757B12EA15.

Rezai, A. R., B. H. Kopell, R. E. Gross, J. L. Vitek, A. D. Sharan, P. Limousin, and A. L. Benabid. 2006. "Deep brain stimulation for Parkinson's disease: Surgical issues." *Movement Disorders* 21 Suppl 14:S197–218. doi:10.1002/mds.20956.

Rousche, P. J. and R. A. Normann. 1998. "Chronic recording capability of the Utah Intracortical Electrode Array in cat sensory cortex." *Journal of Neuroscience Methods* 82 (1):1–15.

Sudhyadhom, A., I. U. Haq, K. D. Foote, M. S. Okun, and F. J. Bova. 2009. "A high resolution and high contrast MRI for differentiation of subcortical structures for DBS targeting: The Fast Gray Matter Acquisition T1 Inversion Recovery (FGATIR)." *Neuroimage* 47 Suppl 2 (19362595):T44–52. doi:papers2://7964EF6A-BE0A-4F2B-9048-9B5C50D5342A/Paper/p10057.

Thongpang, S., T. J. Richner, S. K. Brodnick, A. Schendel, J. Kim, J. A. Wilson, J. Hippensteel et al. 2011. "A micro-electrocorticography platform and deployment strategies for chronic BCI applications." *Clinical EEG and Neuroscience: Official Journal of the EEG and Clinical Neuroscience Society (ENCS)* 42 (4):259–65. doi:papers2://publication/uuid/8DE8A9A1-8843-4DA2-A6F5-5253D5403B10.

Tokuno, H., I. Tanaka, Y. Umitsu, and Y. Nakamura. 2009. "Stereo Navi 2.0: Software for stereotaxic surgery of the common marmoset (Callithrix jacchus)." *Neuroscience Research* 65 (3):312–15. doi:papers2://publication/doi/10.1016/j.neures.2009.08.004.

Venkatraghavan, L., M. Luciano, and P. Manninen. 2010. "Review article: Anesthetic management of patients undergoing deep brain stimulator insertion." *Anesthesia and Analgesia* 110 (4):1138–45. doi:papers2://publication/doi/10.1213/ANE.0b013e3181d2a782.

Chapter 5

Andrews, R. J. 2010. "Neuromodulation: Advances in the next five years." *Annals of the New York Academy of Sciences* 1199:204–11. doi:10.1111/j.1749-6632.2009.05379.x.

Biran, R., D. C. Martin, and P. A. Tresco. 2005. "Neuronal cell loss accompanies the brain tissue response to chronically implanted silicon microelectrode arrays." *Experimental Neurology* 195 (1):115–26.

Biran, R., D. C. Martin, and P. A. Tresco. 2007. "The brain tissue response to implanted silicon microelectrode arrays is increased when the device is tethered to the skull." *Journal of Biomedical Materials Research—Part A* 82 (1):169–78.

Brett, M. A. C. and A. M. O. Brett. 1993. *Electrochemistry: Principles, Methods and Applications*. New York: Oxford Science.

Cogan, S. F. 2008. "Neural stimulation and recording electrodes." *Annual Review of Biomedical Engineering* 10:275–309. doi:10.1146/annurev.bioeng.10.061807.160518.

Collinger, J. L., B. Wodlinger, J. E. Downey, W. Wang, E. C. Tyler-Kabara, D. J. Weber, A. J. C. McMorland, M. Velliste, M. L. Boninger, and A. B. Schwartz. 2013. "High-performance neuroprosthetic control by an individual with tetraplegia." *Lancet* 381 (9866):557–64. doi:10.1016/S0140-6736(12)61816-9.

Csicsvari, J., D. A. Henze, B. Jamieson, K. D. Harris, A. Sirota, P. Barthó, K. D. Wise, and G. Buzsáki. 2003. "Massively parallel recording of unit and local field potentials with silicon-based electrodes." *Journal of Neurophysiology* 90 (2):1314–23.

Dijkstra, C. D., E. A. Dopp, P. Joling, and G. Kraal. 1985. "The heterogeneity of mononuclear phagocytes in lymphoid organs: Distinct macrophage subpopulations in the rat recognized by monoclonal antibodies ED1, ED2 and ED3." *Immunology* 54 (3): 589–99.

Drake, K. L., K. D. Wise, J. Farraye, D. J. Anderson, and S. L. BeMent. 1988. "Performance of planar multisite microprobes in recording extracellular single-unit intracortical activity." *IEEE Transactions on Biomedical Engineering* 35 (9):719–32.

Edell, D. J., V. V. Toi, V. M. McNeil, and L. D. Clark. 1992. "Factors influencing the biocompatibility of insertable silicon microshafts in cerebral cortex." *IEEE Transactions on Biomedical Engineering* 39 (6):635–43.

Ethier, C., E. R. Oby, M. J. Bauman, and L. E. Miller. 2012. "Restoration of grasp following paralysis through brain-controlled stimulation of muscles." *Nature* 485 (7398):368–71. doi:10.1038/Nature10987.

Frampton, J. P., M. R. Hynd, M. L. Shuler, and W. Shain. 2010. "Effects of glial cells on electrode impedance recorded from neuralprosthetic devices in vitro." *Annals of Biomedical Engineering* 38 (3):1031–47.

Freire, M. A., E. Morya, J. Faber, J. R. Santos, J. S. Guimaraes, N. A. Lemos, K. Sameshima, A. Pereira, S. Ribeiro, and M. A. Nicolelis. 2011. "Comprehensive analysis of tissue preservation and recording quality from chronic multielectrode implants." *PLoS One* 6 (11):e27554. doi:10.1371/journal.pone.0027554.

Geddes, L. A. 1997. "Historical evolution of circuit models for the electrode-electrolyte interface." *Annals of Biomedical Engineering* 25 (1):1–14.

Geddes, L. A. and R. Roeder. 2003. "Criteria for the selection of materials for implanted electrodes." *Annals of Biomedical Engineering* 31 (7):879–90.

Graeber, M. B., F. Lopez-Redondo, E. Ikoma, M. Ishikawa, Y. Imai, K. Nakajima, G. W. Kreutzberg, and S. Kohsaka. 1998. "The microglia/macrophage response in the neonatal rat facial nucleus following axotomy." *Brain Research* 813 (2):241–53.

Hochberg, L. R., D. Bacher, B. Jarosiewicz, N. Y. Masse, J. D. Simeral, J. Vogel, S. Haddadin et al. 2012. "Reach and grasp by people with tetraplegia using a neurally controlled robotic arm." *Nature* 485 (7398):372–5. doi:10.1038/nature11076.

References

Iosa, M., G. Morone, A. Fusco, M. Bragoni, P. Coiro, M. Multari, V. Venturiero, D. De Angelis, L. Pratesi, and S. Paolucci. 2012. "Seven capital devices for the future of stroke rehabilitation." *Stroke Research and Treatment* 2012:187965. doi:10.1155/2012/187965.

Jackson, A. and E. E. Fetz. 2007. "Compact movable microwire array for long-term chronic unit recording in cerebral cortex of primates." *Journal of Neurophysiology* 98 (5):3109–18.

Jackson, A. and J. B. Zimmermann. 2012. "Neural interfaces for the brain and spinal cord-restoring motor function." *Nature Reviews Neurology* 8 (12):690–9. doi:10.1038/nrneurol.2012.219.

Johnson, M. D., K. J. Otto, and D. R. Kipke. 2005. "Repeated voltage biasing improves unit recordings by reducing resistive tissue impedances." *IEEE Transactions on Neural Systems and Rehabilitation Engineering* 13 (2):160–5. doi:10.1109/Tnsre.2005.847373.

Kam, L., W. Shain, J. N. Turner, and R. Bizios. 1999. "Correlation of astroglial cell function on micro-patterned surfaces with specific geometric parameters." *Biomaterials* 20 (23–24):2343–50.

Kane, S., S. Cogan, J. Plante, J. Ehrlich, D. McCreery, and P. Troyk. 2013. "Electrical performance of penetrating microelectrodes chronically implanted in cat cortex." *IEEE Transactions on Biomedical Engineering* 60 (8):2153–60. doi:10.1109/TBME.2013.2248152.

Karumbaiah, L., T. Saxena, D. Carlson, K. Patil, R. Patkar, E. A. Gaupp, M. Betancur, G. B. Stanley, L. Carin, and R. V. Bellamkonda. 2013. "Relationship between intracortical electrode design and chronic recording function." *Biomaterials* 34 (33):8061–74. doi:10.1016/j.biomaterials.2013.07.016.

Kim, B. J., J. T. Kuo, S. A. Hara, C. D. Lee, L. Yu, C. A. Gutierrez, T. Q. Hoang, V. Pikov, and E. Meng. 2013. "3D Parylene sheath neural probe for chronic recordings." *Journal of Neural Engineering* 10 (4):045002. doi:10.1088/1741-2560/10/4/045002.

Kipke, D. R., R. J. Vetter, J. C. Williams, and J. F. Hetke. 2003. "Silicon-substrate intracortical microelectrode arrays for long-term recording of neuronal spike activity in cerebral cortex." *IEEE Transactions on Neural Systems and Rehabilitation Engineering* 11 (2):151–5. doi:10.1109/Tnsre.2003.814443.

Koivuniemi, A. S. and K. J. Otto. 2011. "Asymmetric versus symmetric pulses for cortical microstimulation." *IEEE Transactions on Neural Systems and Rehabilitation Engineering* 19 (5):468–76. doi:10.1109/TNSRE.2011.2166563.

Kozai, T. D. Y., N. B. Langhals, P. R. Patel, X. P. Deng, H. N. Zhang, K. L. Smith, J. Lahann, N. A. Kotov, and D. R. Kipke. 2012a. "Ultrasmall implantable composite microelectrodes with bioactive surfaces for chronic neural interfaces." *Nature Materials* 11 (12):1065–73. doi:10.1038/Nmat3468.

Kozai, T. D. Y., A. L. Vazquez, C. L. Weaver, S. G. Kim, and X. T. Cui. 2012b. "In vivo two-photon microscopy reveals immediate microglial reaction to implantation of microelectrode through extension of processes." *Journal of Neural Engineering* 9 (6):066001. doi:10.1088/1741-2560/9/6/066001.

Lee, H., R. V. Bellamkonda, W. Sun, and M. E. Levenston. 2005. "Biomechanical analysis of silicon microelectrode-induced strain in the brain." *Journal of Neural Engineering* 2 (4):81–9.

Lempka, S. F., M. D. Johnson, M. A. Moffitt, K. J. Otto, D. R. Kipke, and C. C. McIntyre. 2011. "Theoretical analysis of intracortical microelectrode recordings." *Journal of Neural Engineering* 8 (4):045006. doi:10.1088/1741-2560/8/4/045006.

Lewicki, M. S. 1998. "A review of methods for spike sorting: The detection and classification of neural action potentials." *Network: Computation in Neural Systems* 9 (4):R53–78.

Maling, N., R. Hashemiyoon, K. D. Foote, M. S. Okun, and J. C. Sanchez. 2012. "Increased thalamic gamma band activity correlates with symptom relief following deep brain stimulation in humans with Tourette's syndrome." *PLoS One* 7 (9):e44215. doi:10.1371/journal.pone.0044215.

McConnell, G. C., H. D. Rees, A. I. Levey, C. A. Gutekunst, R. E. Gross, and R. V. Bellamkonda. 2009. "Implanted neural electrodes cause chronic, local inflammation that is correlated with local neurodegeneration." *Journal of Neural Engineering* 6 (5):056003.

Merrill, D. R., M. Bikson, and J. G. Jefferys. 2005. "Electrical stimulation of excitable tissue: Design of efficacious and safe protocols." *Journal of Neuroscience Methods* 141 (2):171–98.

Morrell, M. J. 2011. "Responsive cortical stimulation for the treatment of medically intractable partial epilepsy." *Neurology* 77 (13):1295–304. doi:10.1212/WNL.0b013e3182302056.

Musallam, S., M. J. Bak, P. R. Troyk, and R. A. Andersen. 2007. "A floating metal microelectrode array for chronic implantation." *Journal of Neuroscience Methods* 160 (1):122–7. doi:10.1016/j.jneumeth.2006.09.005.

Nicolelis, M. A. L., D. Dimitrov, J. M. Carmena, R. Crist, G. Lehew, J. D. Kralik, and S. P. Wise. 2003. "Chronic, multisite, multielectrode recordings in macaque monkeys." *Proceedings of the National Academy of Sciences of the United States of America* 100 (19):11041–6.

Otto, K. J., M. D. Johnson, and D. R. Kipke. 2006. "Voltage pulses change neural interface properties and improve unit recordings with chronically implanted microelectrodes." *IEEE Transactions on Biomedical Engineering* 53 (2):333–40.

Otto, K. J., P. J. Rousche, and D. R. Kipke. 2005. "Cortical microstimulation in auditory cortex of rat elicits best-frequency dependent behaviors." *Journal of Neural Engineering* 2 (2):42–51. doi:10.1088/1741-2560/2/2/005.

Patrick, E., M. E. Orazem, J. C. Sanchez, and T. Nishida. 2011. "Corrosion of tungsten microelectrodes used in neural recording applications." *Journal of Neuroscience Methods* 198 (2):158–71.

Patrick, E., M. Ordonez, N. Alba, J. C. Sanchez, and T. Nishida. 2006. Design and fabrication of a flexible substrate mieroelectrode array for brain machine interfaces. Paper presented at the Annual International Conference of the IEEE Engineering in Medicine and Biology—Proceedings.

Pegg, C. C., C. He, A. R. Stroink, K. A. Kattner, and C. X. Wang. 2010. "Technique for collection of cerebrospinal fluid from the cisterna magna in rat." *Journal of Neuroscience Methods* 187 (1):8–12.

Polikov, V. S., P. A. Tresco, and W. M. Reichert. 2005. "Response of brain tissue to chronically implanted neural electrodes." *Journal of Neuroscience Methods* 148 (1):1–18. doi:papers://7964EF6A-BE0A-4F2B-9048-9B5C50D5342A/Paper/p555.

Prasad, A. and J. C. Sanchez. 2012. "Quantifying long-term microelectrode array functionality using chronic in vivo impedance testing." *Journal of Neural Engineering* 9 (2):026028. doi:10.1088/1741-2560/9/2/026028.

Prasad, A., V. Sankar, A. T. Dyer, E. Knott, Q. S. Xue, T. Nishida, J. R. Reynolds, G. Shaw, W. Streit, and J. C. Sanchez. 2011. "Coupling biotic and abiotic metrics to create a testbed for predicting neural electrode performance." *Conference Proceedings: Annual International Conference of the IEEE Engineering in Medicine and Biology Society* 2011:3020–3. doi:10.1109/IEMBS.2011.6090827.

Prasad, A., Q. S. Xue, V. Sankar, T. Nishida, G. Shaw, W. J. Streit, and J. C. Sanchez. 2012. "Comprehensive characterization and failure modes of tungsten microwire arrays in chronic neural implants." *Journal of Neural Engineering* 9 (5):056015. doi:10.1088/1741-2560/9/5/056015.

Rousche, P. J. and R. A. Normann. 1998. "Chronic recording capability of the Utah Intracortical Electrode Array in cat sensory cortex." *Journal of Neuroscience Methods* 82 (1):1–15.

Rousche, P. J., D. S. Pellinen, D. P. Pivin, Jr., J. C. Williams, R. J. Vetter, and D. R. Kipke. 2001. "Flexible polyimide-based intracortical electrode arrays with bioactive capability." *IEEE Transactions on Biomedical Engineering* 48 (3):361–71. doi:10.1109/10.914800.

References

Rouse, A. G., S. R. Stanslaski, P. Cong, R. M. Jensen, P. Afshar, D. Ullestad, R. Gupta, G. F. Molnar, D. W. Moran, and T. J. Denison. 2011. "A chronic generalized bi-directional brain–machine interface." *Journal of Neural Engineering* 8 (3):036018. doi:10.1088/1741-2560/8/3/036018.

Sanchez, J. C., N. Alba, T. Nishida, C. Batich, and P. R. Carney. 2006. "Structural modifications in chronic microwire electrodes for cortical neuroprosthetics: A case study." *IEEE Transactions on Neural Systems and Rehabilitation Engineering* 14 (2):217–21.

Saxena, T., L. Karumbaiah, E. A. Gaupp, R. Patkar, K. Patil, M. Betancur, G. B. Stanley, and R. V. Bellamkonda. 2013. "The impact of chronic blood-brain barrier breach on intracortical electrode function." *Biomaterials* 34 (20):4703–13. doi:10.1016/J .Biomaterials.2013.03.007.

Schmidt, E. M., M. J. Bak, and J. S. McIntosh. 1976. "Long term chronic recording from cortical neurons." *Experimental Neurology* 52 (3):496–506.

Stensaas, S. S. and L. J. Stensaas. 1978. "Histopathological evaluation of materials implanted in the cerebral cortex." *Acta Neuropathologica* 41 (2):145–55.

Streit, W., Q. S. Xue, A. Prasad, V. Sankar, E. Knott, A. Dyer, J. Reynolds, T. Nishida, G. Shaw, and J. Sanchez. 2012. "Electrode failure: Tissue, electrical, and material responses." *IEEE Pulse* 3 (1):30–3. doi:10.1109/MPUL.2011.2175632.

Streit, W. J., H. Braak, Q. S. Xue, and I. Bechmann. 2009. "Dystrophic (senescent) rather than activated microglial cells are associated with tau pathology and likely precede neuro-degeneration in Alzheimer's disease." *Acta Neuropathologica* 118 (4):475–85.

Suner, S., M. R. Fellows, C. Vargas-Irwin, G. K. Nakata, and J. P. Donoghue. 2005. "Reliability of signals from a chronically implanted, silicon-based electrode array in non-human primate primary motor cortex." *IEEE Transactions on Neural Systems and Rehabilitation Engineering* 13 (4):524–41.

Szarowski, D. H., M. D. Andersen, S. Retterer, A. J. Spence, M. Isaacson, H. G. Craighead, J. N. Turner, and W. Shain. 2003. "Brain responses to micro-machined silicon devices." *Brain Research* 983 (1–2):23–35.

Thelin, J., H. Jorntell, E. Psouni, M. Garwicz, J. Schouenborg, N. Danielsen, and C. E. Linsmeier. 2011. "Implant size and fixation mode strongly influence tissue reactions in the CNS." *PLoS One* 6 (1):e16267. doi:10.1371/journal.pone.0016267.

Truccolo, W., J. A. Donoghue, L. R. Hochberg, E. N. Eskandar, J. R. Madsen, W. S. Anderson, E. N. Brown, E. Halgren, and S. S. Cash. 2011. "Single-neuron dynamics in human focal epilepsy." *Nature Neuroscience* 14 (5):635–41. doi:10.1038/nn.2782.

Turner, J. N., W. Shain, D. H. Szarowski, M. Andersen, S. Martins, M. Isaacson, and H. Craighead. 1999. "Cerebral astrocyte response to micromachined silicon implants." *Experimental Neurology* 156 (1):33–49.

Vetter, R. J., J. C. Williams, J. F. Hetke, E. A. Nunamaker, and D. R. Kipke. 2004. "Chronic neural recording using silicon-substrate microelectrode arrays implanted in cerebral cortex." *IEEE Transactions on Biomedical Engineering* 51 (6):896–904.

Ward, M. P., P. Rajdev, C. Ellison, and P. P. Irazoqui. 2009. "Toward a comparison of micro-electrodes for acute and chronic recordings." *Brain Research* 1282:183–200. doi:papers2 ://publication/doi/10.1016/j.brainres.2009.05.052.

Webster, J. G. 1997. *Medical Instrumentation: Application and Design*, 3rd ed. New York: Wiley.

Williams, J. C., J. A. Hippensteel, J. Dilgen, W. Shain, and D. R. Kipke. 2007. "Complex impedance spectroscopy for monitoring tissue responses to inserted neural implants." *Journal of Neural Engineering* 4 (4):410–23.

Williams, J. C., R. L. Rennaker, and D. R. Kipke. 1999a. "Long-term neural recording characteristics of wire microelectrode arrays implanted in cerebral cortex." *Brain Research Protocols* 4 (3):303–13.

Williams, J. C., R. L. Rennaker, and D. R. Kipke.. 1999b. "Stability of chronic multichannel neural recordings: Implications for a long-term neural interface." *Neurocomputing* 26–7:1069–76.

Winslow, B. D. and P. A. Tresco. 2010. "Quantitative analysis of the tissue response to chronically implanted microwire electrodes in rat cortex." *Biomaterials* 31 (7):1558–67.

Chapter 6

Akaike, H. 1974. "A new look at the statistical model identification." *IEEE Transactions on Automatic Control* 19:716–23.

Bauer, R. and A. Gharabaghi. 2015. "Reinforcement learning for adaptive threshold control of restorative brain–computer interfaces: A Bayesian simulation." *Frontiers in Neuroscience* 9:36. doi:10.3389/fnins.2015.00036.

Brown, E. N., R. E. Kass, and P. P. Mitra. 2004. "Multiple neural spike train data analysis: State-of-the-art and future challenges." *Nature Neuroscience* 7 (5):456–61.

Brunner, P., A. L. Ritaccio, J. F. Emrich, H. Bischof, and G. Schalk. 2011. "Rapid communication with a "P300" matrix speller using electrocorticographic signals (ECoG)." *Frontiers in Neuroscience* 5:5. doi:10.3389/fnins.2011.00005.

Buzsáki, G. 2006. *Rhythms of the Brain.* New York: Oxford University Press.

Calvin, W. H. 1990. "The emergence of intelligence." *Scientific American* 9 (4):44–51.

Carmena, J. M., M. A. Lebedev, R. E. Crist, J. E. O'Doherty, D. M. Santucci, D. Dimitrov, P. G. Patil, C. S. Henriquez, and M. A. Nicolelis. 2003. "Learning to control a brain–machine interface for reaching and grasping by primates." *PLoS Biol* 1 (2):193–208.

Carmena, J. M., M. A. Lebedev, C. S. Henriquez, and M. A. Nicolelis. 2005. "Stable ensemble performance with single-neuron variability during reaching movements in primates." *Journal of Neuroscience* 25 (46):10712–16. doi:10.1523/JNEUROSCI.2772-05.2005.

Chapin, J. K., K. A. Moxon, R. S. Markowitz, and M. A. Nicolelis. 1999. "Real-time control of a robot arm using simultaneously recorded neurons in the motor cortex." *Nature Neuroscience* 2 (7):664–70.

Chase, S. M., R. E. Kass, and A. B. Schwartz. 2012. "Behavioral and neural correlates of visuomotor adaptation observed through a brain–computer interface in primary motor cortex." *Journal of Neurophysiology* 108 (2):624–44. doi:10.1152/jn.00371.2011.

Chestek, C. A., V. Gilja, P. Nuyujukian, J. D. Foster, J. M. Fan, M. T. Kaufman, M. M. Churchland et al. 2011. "Long-term stability of neural prosthetic control signals from silicon cortical arrays in rhesus macaque motor cortex." *Journal of Neural Engineering* 8 (4):045005. doi:10.1088/1741-2560/8/4/045005.

Collinger, J. L., B. Wodlinger, J. E. Downey, W. Wang, E. C. Tyler-Kabara, D. J. Weber, A. J. McMorland, M. Velliste, M. L. Boninger, and A. B. Schwartz. 2013. "High-performance neuroprosthetic control by an individual with tetraplegia." *Lancet* 381 (9866):557–64. doi:10.1016/S0140-6736(12)61816-9.

Dickey, A. S., A. Suminski, Y. Amit, and N. G. Hatsopoulos. 2009. "Single-unit stability using chronically implanted multielectrode arrays." *Journal of Neurophysiology* 102 (2):1331–9. doi:10.1152/jn.90920.2008.

DiGiovanna, J., B. Mahmoudi, J. Fortes, J. C. Principe, and J. C. Sanchez. 2009. "Co-adaptive brain machine interface via reinforcement learning." *IEEE Transactions on Biomedical Engineering (Special Issue on Hybrid Bionics)* 56 (1):54–64. doi:10.1109/TBME.2008.926699.

Eden, U. T., L. M. Frank, R. Barbieri, V. Solo, and E. N. Brown. 2004. "Dynamic analysis of neural encoding by point process adaptive filtering." *Neural Computation* 16:971–98.

References

Fraser, G. W. and A. B. Schwartz. 2012. "Recording from the same neurons chronically in motor cortex." *Journal of Neurophysiology* 107 (7):1970–8. doi:10.1152/jn.01012.2010.

Fraser, G. W., S. M. Chase, A. Whitford, and A. B. Schwartz. 2009. "Control of a brain–computer interface without spike sorting." *Journal of Neural Engineering* 6 (5):055004. doi:10.1088/1741-2560/6/5/055004.

Fuster, J. M. 2004. "Upper processing stages of the perception-action cycle." *Trends in Cognitive Sciences* 8 (4):143–5. doi:10.1016/j.tics.2004.02.004.

Ganguly, K. and J. M. Carmena. 2009. "Emergence of a stable cortical map for neuroprosthetic control." *PLoS Biol* 7 (7):e1000153. doi:10.1371/journal.pbio.1000153.

Ganguly, K., D. F. Dimitrov, J. D. Wallis, and J. M. Carmena. 2011. "Reversible large-scale modification of cortical networks during neuroprosthetic control." *Nature Neuroscience* 14 (5):662–7. doi:10.1038/nn.2797.

Gao, Y., M. J. Black, E. Bienenstock, W. Wu, and J. P. Donoghue. 2003. A quantitative comparison of linear and non-linear models of motor cortical activity for the encoding and decoding of arm motions. Paper presented at the the 1st International IEEE EMBS Conference on Neural Engineering, Capri, Italy.

Geman, S., E. Bienenstock, and R. Doursat. 1992. "Neural networks and the bias/variance dilemma." *Neural Computation* 4:1–58.

Geng, S., N. W. Prins, E. A. Pohlmeyer, A. Prasad, and J. C. Sanchez. 2013. "Extraction of error related local field potentials from the striatum during environmental perturbations of a robotic arm." In *6th International IEEE EMBS Conference on Neural Engineering*, pp. 993–6. doi:10.1109/NER.2013.6696103.

Gilja, V., P. Nuyujukian, C. A. Chestek, J. P. Cunningham, B. M. Yu, J. M. Fan, M. M. Churchland et al. 2012. "A high-performance neural prosthesis enabled by control algorithm design." *Nature Neuroscience* 15 (12):1752–7. doi:10.1038/nn.3265.

Haykin, S. O. 1994. *Neural Networks: A Comprehensive Foundation*. New York; Toronto: Macmillan; Maxwell Macmillan Canada.

Haykin, S. O. 1996. *Adaptive Filter Theory*, 3rd ed. Upper Saddle River, NJ: Prentice-Hall International.

He, B., J. Lian, K. M. Spencer, J. Dien, and E. Donchin. 2001. "A cortical potential imaging analysis of the P300 and novelty P3 components." *Human Brain Mapping* 12 (2):120–30.

Hebb, D. O. 1949. *The Organization of Behavior: A Neuropsychological Theory*. New York: Wiley.

Helms Tillery, S. I., D. M. Taylor, and A. B. Schwartz. 2003. "Training in cortical control of neuroprosthetic devices improves signal extraction from small neuronal ensembles." *Reviews in the Neurosciences* 14:107–19.

Hochberg, L. R., D. Bacher, B. Jarosiewicz, N. Y. Masse, J. D. Simeral, J. Vogel, S. Haddadin et al. 2012. "Reach and grasp by people with tetraplegia using a neurally controlled robotic arm." *Nature* 485 (7398):372–5. doi:10.1038/nature11076.

Hochberg, L. R., M. D. Serruya, G. M. Friehs, J. A. Mukand, M. Saleh, A. H. Caplan, A. Branner, D. Chen, R. D. Penn, and J. P. Donoghue. 2006. "Neuronal ensemble control of prosthetic devices by a human with tetraplegia." *Nature* 442 (7099):164–71.

Hoerl, A. E. and R. W. Kennard. 1970. "Ridge regression: Biased estimation for nonorthogonal problems." *Technometrics* 12 (3):55–67.

Hoffmann, S. and M. Falkenstein. 2012. "Predictive information processing in the brain: Errors and response monitoring." *International Journal of Psychophysiology* 83 (2):208–12. doi:10.1016/j.ijpsycho.2011.11.015.

Iturrate, I., L. Montesano, and J. Minguez. 2010. "Robot reinforcement learning using EEG-based reward signals." In *2010 IEEE International Conference on Robotics and Automation (Icra)*, 4822–9. doi:10.1109/Robot.2010.5509734.

Jarosiewicz, B., S. M. Chase, G. W. Fraser, M. Velliste, R. E. Kass, and A. B. Schwartz. 2008. "Functional network reorganization during learning in a brain–computer interface paradigm." *Proceedings of the National Academy of Sciences of the United States of America* 105 (49):19486–91. doi:10.1073/pnas.0808113105.

Kim, S. P., J. C. Sanchez, Y. N. Rao, D. Erdogmus, J. C. Principe, J. M. Carmena, M. A. Lebedev, and M. A. L. Nicolelis. 2006. "A comparison of optimal MIMO linear and nonlinear models for brain–machine interfaces." *Journal of Neural Engineering* 3 (2):145–61.

Kreilinger, A., C. Neuper, and G. R. Muller-Putz. 2012. "Error potential detection during continuous movement of an artificial arm controlled by brain–computer interface." *Medical & Biological Engineering & Computing* 50 (3):223–30. doi:10.1007/s11517-011-0858-4.

Lebedev, M. A., J. M. Carmena, J. E. O'Doherty, M. Zacksenhouse, C. S. Henriquez, J. C. Principe, and M. A. Nicolelis. 2005. "Cortical ensemble adaptation to represent velocity of an artificial actuator controlled by a brain–machine interface." *Journal of Neuroscience* 25 (19):4681–93. doi:10.1523/JNEUROSCI.4088-04.2005.

Li, Z., J. E. O'Doherty, T. L. Hanson, M. A. Lebedev, C. S. Henriquez, and M. A. Nicolelis. 2009. "Unscented Kalman filter for brain–machine interfaces." *PLoS One* 4 (7):e6243. doi:10.1371/journal.pone.0006243.

Li, Z., J. E. O'Doherty, M. A. Lebedev, and M. A. Nicolelis. 2011. "Adaptive decoding for brain–machine interfaces through Bayesian parameter updates." *Neural Computation* 23 (12):3162–204. doi:10.1162/NECO_a_00207.

Llera, A., M. A. van Gerven, V. Gomez, O. Jensen, and H. J. Kappen. 2011. "On the use of interaction error potentials for adaptive brain computer interfaces." *Neural Networks* 24 (10):1120–7. doi:10.1016/j.neunet.2011.05.006.

Ludwig, K. A., R. M. Miriani, N. B. Langhals, M. D. Joseph, D. J. Anderson, and D. R. Kipke. 2009. "Using a common average reference to improve cortical neuron recordings from microelectrode arrays." *Journal of Neurophysiology* 101 (3):1679–89. doi:10.1152/jn.90989.2008.

Mahmoudi, B. and J. C. Sanchez. 2011. "A symbiotic brain–machine interface through value-based decision making." *PLoS One* 6 (3):e14760. doi:10.1371/journal.pone.0014760.

Mahmoudi, B., E. A. Pohlmeyer, N. W. Prins, S. Geng, and J. C. Sanchez. 2013. "Towards autonomous neuroprosthetic control using Hebbian reinforcement learning." *Journal of Neural Engineering* 10 (6):066005. doi:10.1088/1741-2560/10/6/066005.

Matsuzaki, S., Y. Shiina, and Y. Wada. 2011. "Adaptive classification for brain–machine interface with reinforcement learning." *Neural Information Processing, Pt I* 7062:360–9.

Mehring, C., M. P. Nawrot, S. C. de Oliveira, E. Vaadia, A. Schulze-Bonhage, A. Aertsen, and T. Ball. 2004. "Comparing information about arm movement direction in single channels of local and epicortical field potentials from monkey and human motor cortex." *Journal of Physiology, Paris* 98 (4–6):498–506. doi:10.1016/j.jphysparis.2005.09.016.

Mehring, C., J. Rickert, E. Vaadia, S. Cardosa de Oliveira, A. Aertsen, and S. Rotter. 2003. "Inference of hand movements from local field potentials in monkey motor cortex." *Nature Neuroscience* 6 (12):1253–4. doi:10.1038/nn1158.

Milekovic, T., T. Ball, A. Schulze-Bonhage, A. Aertsen, and C. Mehring. 2012. "Error-related electrocorticographic activity in humans during continuous movements." *Journal of Neural Engineering* 9 (2):026007. doi:10.1088/1741-2560/9/2/026007.

Moran, D. W. and A. B. Schwartz. 1999. "Motor cortical representation of speed and direction during reaching." *Journal of Neurophysiology* 82 (5):2676–92.

Neal, R. 1996. *Bayesian Learning for Neural Networks*. Cambridge: Cambridge University Press.

213

References

Oppenheim, A. V., R. W. Schafer, and J. R. Buck. 1999. *Discrete-Time Signal Processing.* New York: Prentice-Hall.

Orsborn, A. L., S. Dangi, H. G. Moorman, and J. M. Carmena. 2012. "Closed-loop decoder adaptation on intermediate time-scales facilitates rapid BMI performance improvements independent of decoder initialization conditions." *IEEE Transactions on Neural Systems and Rehabilitation Engineering* 20 (4):468–77. doi:10.1109/TNSRE.2012.2185066.

Paninski, L., M. R. Fellows, N. G. Hatsopoulos, and J. P. Donoghue. 2004. "Spatiotemporal tuning of motor cortical neurons for hand position and velocity." *Journal of Neurophysiology* 91 (1):515–32.

Pilarski, P. M., M. R. Dawson, T. Degris, F. Fahimi, J. P. Carey, and R. S. Sutton. 2011. "Online human training of a myoelectric prosthesis controller via actor-critic reinforcement learning." *IEEE International Conference on Rehabilitation Robotics* 2011:1–7. doi:10.1109/ICORR.2011.5975338.

Pohlmeyer, E. A., E. R. Oby, E. J. Perreault, S. A. Solla, K. L. Kilgore, R. F. Kirsch, and L. E. Miller. 2009. "Toward the restoration of hand use to a paralyzed monkey: Brain-controlled functional electrical stimulation of forearm muscles." *PLoS One* 4 (6):e5924. doi:10.1371/journal.pone.0005924.

Pohlmeyer, E. A., S. A. Solla, E. J. Perreault, and L. E. Miller. 2007. "Prediction of upper limb muscle activity from motor cortical discharge during reaching." *Journal of Neural Engineering* 4 (4):369–79. doi:10.1088/1741-2560/4/4/003.

Pohlmeyer, E. A., B. Mahmoudi, S. Geng, N. W. Prins, and J. C. Sanchez. 2012. "Brain–machine interface control of a robot arm using actor-critic reinforcement learning." *Conference Proceedings: Annual International Conference of the IEEE Engineering in Medicine and Biology Society* 2012:4108–11. doi:10.1109/EMBC.2012.6346870.

Pohlmeyer, E. A., B. Mahmoudi, S. Geng, N. W. Prins, and J. C. Sanchez. 2014. "Using reinforcement learning to provide stable brain–machine interface control despite neural input reorganization." *PLoS One* 9 (1):e87253. doi:10.1371/journal.pone.0087253.

Prasad, A., Q. S. Xue, V. Sankar, T. Nishida, G. Shaw, W. J. Streit, and J. C. Sanchez. 2012. "Comprehensive characterization and failure modes of tungsten microwire arrays in chronic neural implants." *Journal of Neural Engineering* 9 (5):056015. doi:10.1088/1741-2560/9/5/056015.

Prins, N. W., S. Geng, E. A. Pohlmeyer, B. Mahmoudi, and J. C. Sanchez. 2013. "Feature extraction and unsupervised classification of neural population reward signals for reinforcement based BMI." In *Engineering in Medicine and Biology Society (EMBC), 2013 Annual International Conference of the IEEE.*

Sanchez, J. C. and J. C. Principe. 2006. "Optimal signal processing for brain–machine interfaces." In *Handbook of Neural Engineering*, edited by M. Akay. New York: Wiley.

Sanchez, J. C. and J. C. Principe. 2007. *Brain Machine Interface Engineering.* San Rafael, CA: Morgan and Claypool.

Sanchez, J. C., S. P. Kim, D. Erdogmus, Y. N. Rao, J. C. Principe, J. Wessberg, and M. A. L. Nicolelis. 2002. Input–output mapping performance of linear and nonlinear models for estimating hand trajectories from cortical neuronal firing patterns. Paper presented at the International Work on Neural Networks for Signal Processing, Martigny, Switzerland, September.

Sanchez, J. C., J. M. Carmena, M. A. Lebedev, M. A. L. Nicolelis, J. G. Harris, and J. C. Principe. 2004. "Ascertaining the importance of neurons to develop better brain-machine interfaces." *IEEE Transactions on Biomedical Engineering* 51 (6):943–53.

Santhanam, G., M. D. Linderman, V. Gilja, A. Afshar, S. I. Ryu, T. H. Meng, and K. V. Shenoy. 2007. "HermesB: A continuous neural recording system for freely behaving primates." *IEEE Transactions on Biomedical Engineering* 54 (11):2037–50. doi:10.1109/TBME.2007.895753.

Santhanam, G., S. I. Ryu, B. M. Yu, A. Afshar, and K. V. Shenoy. 2006. "A high-performance brain–computer interface." *Nature* 442 (7099):195–8. doi:10.1038/nature04968.

Schultz, W. 2000. "Multiple reward signals in the brain." *Nature Reviews Neuroscience* 1 (3):199–207. doi:10.1038/35044563.

Serruya, M. D., N. G. Hatsopoulos, L. Paninski, M. R. Fellows, and J. P. Donoghue. 2002. "Brain–machine interface: Instant neural control of a movement signal." *Nature* 416 (6877):141–2.

Simeral, J. D., S. P. Kim, M. J. Black, J. P. Donoghue, and L. R. Hochberg. 2011. "Neural control of cursor trajectory and click by a human with tetraplegia 1000 days after implant of an intracortical microelectrode array." *Journal of Neural Engineering* 8 (2):025027. doi:10.1088/1741-2560/8/2/025027.

Spuler, M., W. Rosenstiel, and M. Bogdan. 2012. "Online adaptation of a c-VEP Brain–Computer Interface (BCI) based on error-related potentials and unsupervised learning." *PLoS One* 7 (12):e51077. doi:10.1371/journal.pone.0051077.

Sussillo, D., P. Nuyujukian, J. M. Fan, J. C. Kao, S. D. Stavisky, S. Ryu, and K. Shenoy. 2012. "A recurrent neural network for closed-loop intracortical brain–machine interface decoders." *Journal of Neural Engineering* 9 (2):026027. doi:10.1088/1741-2560/9/2/026027.

Sutton, R. S. and A. G. Barto. 1998. *Reinforcement Learning: An Introduction, Adaptive Computation and Machine Learning.* Cambridge, MA: MIT Press.

Taylor, D. M., S. I. Tillery, and A. B. Schwartz. 2002. "Direct cortical control of 3D neuroprosthetic devices." *Science* 296 (5574):1829–32.

Truccolo, W., U. T. Eden, M. R. Fellows, J. P. Donoghue, and E. N. Brown. 2004. "Innovative methodology a point process framework for relating neural spiking activity to spiking history, neural ensemble, and extrinsic covariate effects." *Journal of Neurophysiology* 93:1074–89.

Turner, J. N., W. Shain, D. H. Szarowski, M. Andersen, S. Martins, M. Isaacson, and H. Craighead. 1999. "Cerebral astrocyte response to micromachined silicon implants." *Experimental Neurology* 156 (1):33–49. doi:10.1006/exnr.1998.6983.

Vapnik, V. 1999. *The Nature of Statistical Learning Theory, Statistics for Engineering and Information Science.* New York: Springer-Verlag.

Velliste, M., S. Perel, M. C. Spalding, A. S. Whitford, and A. B. Schwartz. 2008. "Cortical control of a prosthetic arm for self-feeding." *Nature* 453 (7198):1098–101. doi:10.1038/nature06996.

Wahba, G. 1990. *Spline Models for Observational Data.* Montpelier: Capital City Press.

Wang, W., A. D. Degenhart, J. L. Collinger, R. Vinjamuri, G. P. Sudre, P. D. Adelson, D. L. Holder et al. 2009. "Human motor cortical activity recorded with Micro-ECoG electrodes, during individual finger movements." *Conference Proceedings: Annual International Conference of the IEEE Engineering in Medicine and Biology Society IEEE Engineering in Medicine and Biology Society Conference* 2009:586–9. doi:10.1109/IEMBS.2009.5333704.

Wessberg, J., C. R. Stambaugh, J. D. Kralik, P. D. Beck, M. Laubach, J. K. Chapin, J. Kim, S. J. Biggs, M. A. Srinivasan, and M. A. L. Nicolelis. 2000. "Real-time prediction of hand trajectory by ensembles of cortical neurons in primates." *Nature* 408 (6810):361–5.

Wu, W., M. J. Black, Y. Gao, E. Bienenstock, M. Serruya, and J. P. Donoghue. 2002. Inferring hand motion from multi-cell recordings in motor cortex using a Kalman filter. Paper presented at the SAB Workshop on Motor Control in Humans and Robots: On the Interplay of Real Brains and Artificial Devices, University of Edinburgh, Scotland, August.

Chapter 7

Angeli, C. A., V. R. Edgerton, Y. P. Gerasimenko, and S. J. Harkema. 2014. "Altering spinal cord excitability enables voluntary movements after chronic complete paralysis in humans." *Brain* 137 (Pt 5):1394–409. doi:10.1093/brain/awu038.

Edelman, G. M. 2004. *Wider Than the Sky: The Phenomenal Gift of Consciousness*. New Haven, CT: Yale University Press.

Freeman, W. J. 1975. *Mass Action in the Nervous System: Examination of the Neurophysiological Basis of Adaptive Behavior through EEG*. New York: Academic Press.

Gorman, P. H. and J. T. Mortimer. 1983. "The effect of stimulus parameters on the recruitment characteristics of direct nerve stimulation." *IEEE Transactions on Biomedical Engineering* 30:407–14.

Grill, W. M. and J. T. Mortimer. 1995. "Stimulus waveforms for selective neural stimulation." *IEEE Engineering in Medicine and Biology* 14:375–85.

Hampson, R. E., D. Song, I. Opris, L. M. Santos, D. C. Shin, G. A. Gerhardt, V. Z. Marmarelis, T. W. Berger, and S. A. Deadwyler. 2013. "Facilitation of memory encoding in primate hippocampus by a neuroprosthesis that promotes task-specific neural firing." *Journal of Neural Engineering* 10 (6):066013. doi:10.1088/1741-2560/10/6/066013.

Hodgkin, A. and A. Huxley. 1952. "A quantitative description of membrane current and its application to conduction and excitation in nerve." *The Journal of Physiology* 117: 500–44.

Johnson, M. D., K. J. Otto, and D. R. Kipke. 2005. "Repeated voltage biasing improves unit recordings by reducing resistive tissue impedances." *IEEE Transactions on Neural Systems and Rehabilitation Engineering: A Publication of the IEEE Engineering in Medicine and Biology Society* 13 (2):160–5. doi:papers2://publication/doi/10.1109/TNSRE.2005.847373.

Lozano, A. M. and N. Mahant. 2004. "Deep brain stimulation surgery for Parkinson's disease: Mechanisms and consequences." *Pakinsonism and Related Disorders* 10:S49–57.

McIntyre, C. C. 2011. "The electrode—Principles of the neural interface: Axons and cell bodies." In *Essential Neuromodulation*, edited by J. E. Arle and J. L. Shils, 153–68. San Diego, CA: Academic Press.

Merrill, D. R. 2011. "The electrode—Materials and configurations." In *Essential Neuromodulation*, edited by J. E. Arle and J. L. Shils, 107–52. San Diego, CA: Academic Press.

Merrill, D. R., M. Bikson, and J. G. R. Jefferys. 2005. "Electrical stimulation of excitable tissue: Design of efficacious and safe protocols." *Journal of Neuroscience Methods* 141 (2):171–98. doi:10.1016/j.jneumeth.2004.10.020.

Peckham, P. H. and J. S. Knutson. 2005. "Functional electrical stimulation for neuromuscular applications." *Annual Review of Biomedical Engineering* 7:327–60. doi:10.1146/annurev.bioeng.6.040803.140103.

Plonsey, R. 1969. *Bioelectric Phenomena*. New York: McGraw-Hill.

Purves, D., G. J. Augustine, D. Fitzpatrick, L. C. Katz, A.-S. LaMantia, J. O. McNamara, and S. M. Williams. 2001. *Neuroscience*. Sunderland, MA: Sinauer Associates.

Tabot, G. A., S. S. Kim, J. E. Winberry, and S. J. Bensmaia. 2014. "Restoring tactile and proprioceptive sensation through a brain interface." *Neurobiology of Disease*. S0969-9961 (14)00260-5. doi: 10.1016/j.nbd.2014.08.029. [Epub ahead of print].

216

Chapter 8

Anderson, K. D. 2004. "Targeting recovery: Priorities of the spinal cord-injured population." *Journal of Neurotrauma* 21 (10):1371–83.

Anderson, K. D. 2009. "Consideration of user priorities when developing neural prosthetics." *Journal of Neural Engineering* 6 (5):055003. doi:10.1088/1741-2560/6/5/055003.

Ang, K. K., C. Guan, K. S. G. Chua, B.-T. Ang, C. Kuah, C. Wang, K. S. Phua, Z. Y. Chin, and H. Zhang. 2010. Clinical study of neurorehabilitation in stroke using EEG-based motor imagery brain–computer interface with robotic feedback. Paper presented at the Engineering in Medicine and Biology Society (EMBC), 2010 Annual International Conference of the IEEE, August 31–September 4.

Beekhuizen, K. S. and E. C. Field-Fote. 2005. "Massed practice versus massed practice with stimulation: Effects on upper extremity function and cortical plasticity in individuals with incomplete cervical spinal cord injury." *Neurorehabilitation and Neural Repair* 19 (1):33–45. doi:10.1177/1545968305274517.

Beekhuizen, K. S. and E. C. Field-Fote. 2008. "Sensory stimulation augments the effects of massed practice training in persons with tetraplegia." *Archives of Physical Medicine and Rehabilitation* 89 (4):602–8. doi:10.1016/j.apmr.2007.11.021.

Broetz, D., C. Braun, C. Weber, S. R. Soekadar, A. Caria, and N. Birbaumer. 2010. "Combination of brain–computer interface training and goal-directed physical therapy in chronic stroke: a case report." *Neurorehabilitation and Neural Repair* 24 (7):674–9.

Bruehlmeier, M., V. Dietz, K. L. Leenders, U. Roelcke, J. Missimer, and A. Curt. 1998. "How does the human brain deal with a spinal cord injury?" *European Journal of Neuroscience* 10 (12):3918–22.

Bryden, A. M., K. L. Kilgore, B. B. Lind, and D. T. Yu. 2004. "Triceps denervation as a predictor of elbow flexion contractures in C5 and C6 tetraplegia." *Archives of Physical Medicine and Rehabilitation* 85 (11):1880–5. doi:10.1016/J.Apmr.2004.01.042.

Bryden, A. M., W. D. Memberg, and P. E. Crago. 2000. "Electrically stimulated elbow extension in persons with C5/C6 tetraplegia: A functional and physiological evaluation." *Archives of Physical Medicine and Rehabilitation* 81 (1):80–8.

Chiou, Y.-H., J.-J. Luh, S.-C. Chen, Y.-L. Chen, J.-S. Lai, and T.-S. Kuo. 2009. "Patient-driven loop control for hand function restoration in a non-invasive functional electrical stimulation system." *Disability and Rehabilitation* 30 (19):1499–505. doi:10.1080/09638280701615246.

Cramer, S., E. Orr, M. Cohen, and M. Lacourse. 2007. "Effects of motor imagery training after chronic, complete spinal cord injury." *Experimental Brain Research* 177 (2):233–42. doi:10.1007/s00221-006-0662-9.

Cramer, S. C., L. Lastra, M. G. Lacourse, and M. J. Cohen. 2005. "Brain motor system function after chronic, complete spinal cord injury." *Brain* 128 (12):2941–50. doi:10.1093/brain/awh648.

Daly, J. J. and J. R. Wolpaw. 2008. "Brain–computer interfaces in neurological rehabilitation." *The Lancet Neurology* 7 (11):1032–43.

Daly, J. J., R. Cheng, J. Rogers, K. Litinas, K. Hrovat, and M. Dohring. 2009. "Feasibility of a new application of noninvasive Brain Computer Interface (BCI): A case study of training for recovery of volitional motor control after stroke." *Journal of Neurologic Physical Therapy* 33 (4):203–11.

Doucet, B. M., A. Lam, and L. Griffin. 2012. "Neuromuscular electrical stimulation for skeletal muscle function." *The Yale Journal of Biology and Medicine* 85 (2):201–15.

References

Falkenstein, M., J. Hoormann, S. Christ, and J. Hohnsbein. 2000. "ERP components on reaction errors and their functional significance: A tutorial." *Biological Psychology* 51 (2–3):87–107. doi:10.1016/s0301-0511(99)00031-9.

Ferrez, P. W. and J. Millan. 2008. "Error-related EEG potentials generated during simulated brain–computer interaction." *Biomedical Engineering, IEEE Transactions on* 55 (3):923–9.

Green, J. B., E. Sora, Y. Bialy, A. Ricamato, and R. W. Thatcher. 1998. "Cortical sensorimotor reorganization after spinal cord injury—An electroencephalographic study." *Neurology* 50 (4):1115–21.

Hoffman, L. R. and E. C. Field-Fote. 2006. "Cortically-evoked potentials of muscles distal to the lesion are posteriorly shifted and of lower amplitude in individuals with tetraplegia due to spinal cord injury." *Journal of Neurologic Physical Therapy* 30 (4):202–3.

Hoffman, L. R. and E. C. Field-Fote. 2007. "Cortical reorganization following bimanual training and somatosensory stimulation in cervical spinal cord injury: A case report." *Physical Therapy* 87 (2):208–23. doi:10.2522/ptj.20050365.

Hoffman, L. R. and E. C. Field-Fote. 2010. "Functional and corticomotor changes in individuals with tetraplegia following unimanual or bimanual massed practice training with somatosensory stimulation: A pilot study." *Journal of Neurologic Physical Therapy* 34 (4):193–201.

Jebsen, R. H., N. Taylor, R. B. Trieschmann, M. J. Trotter, and L. A. Howard. 1969. "An objective and standardized test of hand function." *Archives of Physical Medicine and Rehabilitation* 50 (6):311–19.

Kilgore, K. L. and P. H. Peckham. 1993. "Grasp synthesis for upper-extremity Fns. 1. Automated-method for synthesizing the stimulus map." *Medical & Biological Engineering & Computing* 31 (6):607–14.

Kilgore, K. L., H. A. Hoyen, A. M. Bryden, R. L. Hart, M. W. Keith, and P. H. Peckham. 2008. "An implanted upper-extremity neuroprosthesis using myoelectric control." *Journal of Hand Surgery-American* 33A (4):539–50. doi:10.1016/J.Jhsa.2008.01.007.

Kirshblum, S. C., S. P. Burns, F. Biering-Sørensen, W. Donovan, D. E. Graves, A. Jha, M. Johansen et al. 2013. "International standards for neurological classification of spinal cord injury (Revised 2011)." *The Journal of Spinal Cord Medicine* 34 (6):535–46. doi:10.1179/204577211X13207446293695.

Kokotilo, K. J., J. J. Eng, and A. Curt. 2009. "Reorganization and preservation of motor control of the brain in spinal cord injury: A systematic review." *Journal of Neurotrauma* 26 (11):2113–26. doi:10.1089/neu.2008.0688.

LeCun, Y., L. Bottou, G. B. Orr, and K.-R. Müller. 1998. "Efficient backprop." In *Neural Networks: Tricks of the Trade*, edited by G. Montavon, G. B. Orr, and K.-R. Müller, 9–50. Berlin: Springer.

Levy, W. J., V. E. Amassian, M. Traad, and J. Cadwell. 1990. "Focal magnetic coil stimulation reveals motor cortical system reorganized in humans after traumatic quadriplegia." *Brain Research* 510 (1):130–4.

Liepert, J., H. Bauder, W. H. R. Miltner, E. Taub, and C. Weiller. 2000. "Treatment-induced cortical reorganization after stroke in humans." *Stroke* 31 (6):1210–16. doi:10.1161/01.str.31.6.1210.

Mahmoudi, B. and J. C. Sanchez. 2011. "A symbiotic brain–machine interface through value-based decision making." *PLoS One* 6 (3):e14760. doi:10.1371/journal.pone.0014760.

Mangold, S., T. Keller, A. Curt, and V. Dietz. 2005. "Transcutaneous functional electrical stimulation for grasping in subjects with cervical spinal cord injury." *Spinal Cord* 43 (1):1–13. doi:10.1038/Sj.Sc.3101644.

Moss, C. W., K. L. Kilgore, and P. H. Peckham. 2011. "A novel command signal for motor neuroprosthetic control." *Neurorehabilitation and Neural Repair* 25 (9):847–54. doi:10.1177/1545968311410067.

218

Mulcahey, M. J., B. T. Smith, and R. R. Betz. 2004. "Psychometric rigor of the grasp and release test for measuring functional limitation of persons with tetraplegia: A preliminary analysis." *The Journal of Spinal Cord Medicine* 27 (1):41–6.

Müller-Putz, G. R., R. Scherer, G. Pfurtscheller, and R. Rupp. 2005. "EEG-based neuroprosthesis control: A step towards clinical practice." *Neuroscience Letters* 382 (1–2):169–74. doi:http://dx.doi.org/10.1016/j.neulet.2005.03.021.

Page, S. J., V. H. Hermann, P. G. Levine, E. Lewis, J. Stein, and J. DePeel. 2011. "Portable neurorobotics for the severely affected arm in chronic stroke: A case study." *Journal of Neurologic Physical Therapy* 35 (1):41–6.

Peckham, P. H. and J. S. Knutson. 2005. "Functional electrical stimulation for neuromuscular applications." *Annual Review of Biomedical Engineering* 7:327–60. doi:10.1146/annurev .bioeng.6.040803.140103.

Peckham, P. H., M. W. Keith, K. L. Kilgore, J. H. Grill, K. S. Wuolle, G. B. Thrope, P. Gorman et al. 2001. "Efficacy of an implanted neuroprosthesis for restoring hand grasp in tetraplegia: A multicenter study." *Archives of Physical Medicine and Rehabilitation* 82 (10):1380–8. doi:10.1053/Apmr.2001.25910.

Pfurtscheller, G., C. Guger, G. Müller, G. Krausz, and C. Neuper. 2000. "Brain oscillations control hand orthosis in a tetraplegic." *Neuroscience Letters* 292 (3):211–14. doi:10.1016/s0304-3940(00)01471-3.

Pfurtscheller, G., G. R. Müller, J. Pfurtscheller, H. J. Gerner, and R. Rupp. 2003. "'Thought'— Control of functional electrical stimulation to restore hand grasp in a patient with tetraplegia." *Neuroscience Letters* 351 (1):33–6. doi:http://dx.doi.org/10.1016/S0304 -3940(03)00947-9.

Pohlmeyer, E. A., B. Mahmoudi, G. Shijia, N. Prins, and J. C. Sanchez. 2012. Brain–machine interface control of a robot arm using actor-critic reinforcement learning. Paper presented at the Engineering in Medicine and Biology Society (EMBC), 2012 Annual International Conference of the IEEE, August 28–September 1.

Popovic, M. R., N. Kapadia, V. Zivanovic, J. C. Furlan, B. C. Craven, and C. McGillivray. 2011. "Functional electrical stimulation therapy of voluntary grasping versus only conventional rehabilitation for patients with subacute incomplete tetraplegia: A randomized clinical trial." *Neurorehabilitation and Neural Repair* 25 (5):433–42. doi:10.1177/1545968310392924.

Prechelt, L. 1998. "Early stopping—But when?" In *Neural Networks: Tricks of the Trade*, edited by G. Orr and K.-R. Müller, 553. Berlin/Heidelberg: Springer.

Qin, L., L. Ding, and B. He. 2005. "Motor imagery classification by means of source analysis for brain–computer interface applications." *Journal of Neural Engineering* 2 (4): 65–72.

Raineteau, O. and M. E. Schwab. 2001. "Plasticity of motor systems after incomplete spinal cord injury." *Nature Reviews Neuroscience* 2 (4):263–73.

Scott, T. R. and M. Haugland. 2001. "Command and control interfaces for advanced neuroprosthetic applications." *Neuromodulation* 4 (4):165–75. doi:10.1046/j.1525-1403.2001.00165.x.

Sollerman, C. and A. Ejeskär. 1995. "Sollerman hand function test. A standardised method and its use in tetraplegic patients." *Scandinavian Journal of Plastic and Reconstructive Surgery and Hand Surgery/Nordisk plastikkirurgisk forening [and] Nordisk klubb for handkirurgi* 29 (2):167–76.

Stroh, K. C. and C. L. Van Doren. 1994. "An ADL test to assess hand function with tetraplegic patients." *Journal of Hand Therapy* 7 (1):47–8.

Sutton, R. S. and A. G. Barto. 1998. *Reinforcement Learning: An Introduction*, Vol. 1. Cambridge MA: Cambridge University Press.

Thomas, C. K. and I. Zijdewind. 2006. "Fatigue of muscles weakened by death of motoneurons." *Muscle & Nerve* 33 (1):21–41. doi:10.1002/mus.20400.

Thomas Mortimer, J. 2011. Motor Prostheses. Handbook of Physiology, The Nervous System, Motor Control. *Comprehensive Physiology Supplement* 2: 155-187. doi: 10.1002/cphy .cp010205.

Várkuti, B., C. Guan, Y. Pan, K. S. Phua, K. K. Ang, C. W. K. Kuah, K. Chua, B. T. Ang, N. Birbaumer, and R. Sitaram. 2013. "Resting state changes in functional connectivity correlate with movement recovery for BCI and robot-assisted upper-extremity training after stroke." *Neurorehabilitation and Neural Repair* 27 (1):53–62.

Wall, P. D. and M. D. Egger. 1971. "Formation of new connexions in adult rat brains after partial deafferentation." *Nature* 232 (5312):542–5.

Wuolle, K. S., C. L. Vandoren, G. B. Thrope, M. W. Keith, and P. H. Peckham. 1994. "Development of a quantitative hand grasp and release test for patients with tetraplegia using a hand neuroprosthesis." *Journal of Hand Surgery-American Volume* 19A (2):209–18.

Chapter 9

Aquilina, O. 2006. "A brief history of cardiac pacing." *Images in Paediatric Cardiology* 8 (2):17–81.

Bashirullah, R. 2010. "Wireless implants." *Microwave Magazine, IEEE* 11 (7):S14–S23. doi:10 .1109/MMM.2010.938579.

Borton, D. A., M. Yin, J. Aceros, and A. Nurmikko. 2013. "An implantable wireless neural interface for recording cortical circuit dynamics in moving primates." *Journal of Neural Engineering* 10 (2):026010. doi:10.1088/1741-2560/10/2/026010.

Chen, S., W. Pei, Q. Gui, Y. Chen, S. Zhao, H. Wang, and H. Chen. 2013. "A fiber-based implantable multi-optrode array with contiguous optical and electrical sites." *Journal of Neural Engineering* 10 (4):046020. doi:10.1088/1741-2560/10/4/046020.

Harrison, R. R., R. J. Kier, C. A. Chestek, V. Gilja, P. Nuyujukian, S. Ryu, B. Greger, F. Solzbacher, and K. V. Shenoy. 2009. "Wireless neural recording with single low-power integrated circuit." *IEEE Transactions on Neural Systems and Rehabilitation Engineering: A Publication of the IEEE Engineering in Medicine and Biology Society* 17 (19497825):322–9. doi:papers://7964EF6A-BE0A-4F2B-9048-9B5C50D5342A/Paper /p7999.

Konrad, P. and T. Shanks. 2010. "Implantable brain computer interface: Challenges to neurotechnology translation." *Neurobiology of Disease* 38 (3):369–75. doi:10.1016/j.nbd.2009 .12.007.

Lewicki, M. S. 1998. "A review of methods for spike sorting: The detection and classification of neural action potentials." *Network: Computation in Neural Systems* 9 (4):R53–78.

Maskooki, A., C. B. Soh, E. Gunawan, and K. S. Low. 2011. "Opportunistic routing for body area network." In *Consumer Communications and Networking Conference (CCNC), 2011 IEEE*, 237–41. IEEE.

Nurmikko, A. V., J. P. Donoghue, L. R. Hochberg, W. R. Patterson, Y.-K. Song, C. W. Bull, D. A. Borton et al. 2010. "Listening to brain microcircuits for interfacing with external world-progress in wireless implantable microelectronic neuroengineering devices: Experimental systems are described for electrical recording in the brain using multiple microelectrodes and short range implantable or wearable broadcasting units." *Proceedings of the IEEE. Institute of Electrical and Electronics Engineers* 98 (3):375–88. doi:10.1109/JPROC.2009.2038949.

220

Patrick, E., V. Sankar, W. Rowe, J. C. Sanchez, and T. Nishida. 2009. "Design of an implantable intracortical microelectrode system for brain–machine interfaces." In *Neural Engineering, 2009. NER '09. 4th International IEEE/EMBS Conference on*, 379–82. doi:10.1109/NER.2009.5109312.

Patrick, E., V. Sankar, W. Rowe, J. C. Sanchez, and T. Nishida. 2010. "An implantable integrated low-power amplifier-microelectrode array for Brain–Machine Interfaces." *Conference Proceedings: Annual International Conference of the IEEE Engineering in Medicine and Biology Society IEEE Engineering in Medicine and Biology Society Conference* 2010:1816–19. doi:10.1109/IEMBS.2010.5626419.

Poppendieck, W., A. Sossalla, M.-O. Krob, C. Welsch, T. A. K. Nguyen, W. Gong, J. Digiovanna, S. Micera, D. M. Merfeld, and K.-P. Hoffmann. 2014. "Development, manufacturing and application of double-sided flexible implantable microelectrodes." *Biomedical Microdevices* 16 (6):837–50. doi:10.1007/s10544-014-9887-8.

Rapoport, B. I., J. T. Kedzierski, and R. Sarpeshkar. 2012. "A glucose fuel cell for implantable brain–machine interfaces." *PLoS One* 7 (6):e38436. doi:10.1371/journal.pone.0038436.

Rouse, A. G., S. R. Stanslaski, P. Cong, R. M. Jensen, P. Afshar, D. Ullestad, R. Gupta, G. F. Molnar, D. W. Moran, and T. J. Denison. 2011. "A chronic generalized bi-directional brain–machine interface." *Journal of Neural Engineering* 8 (3):036018. doi:papers2 ://publication/doi/10.1088/1741-2560/8/3/036018.

Ryapolova-Webb, E., P. Afshar, S. Stanslaski, T. Denison, C. de Hemptinne, K. Bankiewicz, and P. A. Starr. 2014. "Chronic cortical and electromyographic recordings from a fully implantable device: Preclinical experience in a nonhuman primate." *Journal of Neural Engineering* 11 (1):016009. doi:10.1088/1741-2560/11/1/016009.

Sanchez, J. C., J. C. Principe, T. T. Nishida, R. Bashirullah, J. G. Harris, and J. A. B. Fortes. 2008. "Technology and signal processing for brain–machine interfaces." *IEEE Signal Processing Magazine* 25 (1):29–40. doi:10.1109/MSP.2007.909525.

Stanslaski, S., P. Afshar, P. Cong, J. Giftakis, P. Stypulkowski, D. Carlson, D. Linde, D. Ullestad, A.-T. Avestruz, and T. Denison. 2012. "Design and validation of a fully implantable, chronic, closed-loop neuromodulation device with concurrent sensing and stimulation." *IEEE Transactions on Neural Systems and Rehabilitation Engineering: A Publication of the IEEE Engineering in Medicine and Biology Society* 20 (4):410–21. doi:10.1109/TNSRE .2012.2183617.

Stieglitz, T. 2010. "Manufacturing, assembling and packaging of miniaturized neural implants." *Microsystem Technologies* 16 (5):723–34.

Streit, W. J., Q.-S. Xue, A. Prasad, V. Sankar, E. Knott, A. Dyer, J. Reynolds, T. Nishida, G. Shaw, and J. Sanchez. 2012. "Tissue, electrical, and material responses in electrode failure." *IEEE Pulse* 3 (1):30–3.

Wise, K. D., D. J. Anderson, and J. F. Hetke. 2003. "Wireless implantable microsystems: High-density electronic interfaces to the nervous system." In *Proceedings of the IEEE* 92 (1):76–97. doi:10.1109/JPROC.2003.820544.

Wise, K. D., A. M. Sodagar, Y. Yao, M. N. Gulari, G. E. Perlin, and K. Najafi. 2008. "Microelectrodes, microelectronics, and implantable neural microsystems." *Proceedings of the IEEE* 96 (7):1184–202. doi:10.1109/JPROC.2008.922564.

Zhang, F., M. Aghagolzadeh, and K. Oweiss. 2012. "A fully implantable, programmable and multimodal neuroprocessor for wireless, cortically controlled brain–machine interface applications." *Journal of Signal Processing Systems* 69 (3):351–61. doi:10.1007/s11265 -012-0670-x.

Chapter 10

Ackermans, L., Y. Temel, D. Cath, C. van der Linden, R. Bruggeman, M. Kleijer, P. Nederveen et al. 2006. "Deep brain stimulation in Tourette's syndrome: Two targets?" *Movement Disorders* 21 (5):709–13. doi:10.1002/mds.20816.

American Psychiatric Association. 2000. *DSM-IV-TR*. Washington, DC: American Psychiatric Association.

Benabid, A. L. 2010. "[Stimulation therapies for Parkinson's disease: Over the past two decades]." *Bulletin de l'Académie Nationale de Médecine* 194 (7):1273–86.

Berendse, H. W. and H. J. Groenewegen. 1991. "Restricted cortical termination fields of the midline and intralaminar thalamic nuclei in the rat." *Neuroscience* 42 (1):73–102.

Bloch, M. H., J. F. Leckman, H. Zhu, and B. S. Peterson. 2005. "Caudate volumes in childhood predict symptom severity in adults with Tourette syndrome." *Neurology* 65 (8):1253–8. doi:10.1212/01.wnl.0000180957.98702.69.

Bloch, M. H., B. S. Peterson, L. Scahill, J. Otka, L. Katsovich, H. Zhang, and J. F. Leckman. 2006. "Adulthood outcome of tic and obsessive-compulsive symptom severity in children with Tourette syndrome." *Archives of Pediatrics and Adolescent Medicine* 160 (1):65–9. doi:10.1001/archpedi.160.1.65.

Bront-Stewart, H. 2011. "The effect of deep brain stimulation on abnormal neural oscillations and hypersynchronzy in Parkinson's disease and dystonia." In *Dynamical Neuroscience XIX: Deep Brain Stimulation in Mental Illness, Neurological Disorders and Cognitive Impairment.* Washington, DC.

Brovelli, A., J. P. Lachaux, P. Kahane, and D. Boussaoud. 2005. "High gamma frequency oscillatory activity dissociates attention from intention in the human premotor cortex." *Neuroimage* 28 (1):154–64. doi:10.1016/j.neuroimage.2005.05.045.

Buzsáki, G., C. A. Anastassiou, and C. Koch. 2012. "The origin of extracellular fields and currents—EEG, ECoG, LFP and spikes." *Nature Reviews. Neuroscience* 13 (6):407–20. doi:10.1038/nrn3241.

Cavanna, A. E. and A. Nani. 2013. "Tourette syndrome and consciousness of action." *Tremor and Other Hyperkinetic Movements (New York, N.Y.)* 3.

Cavanna, A. E. and S. Seri. 2013. "Tourette's syndrome." *British Medical Journal* 347: f4964.

Cavanna, A. E. and C. Termine. 2012. "Tourette syndrome." *Advances in Experimental Medicine and Biology* 724:375–83. doi:10.1007/978-1-4614-0653-2_28.

Cavanna, A. E., S. Servo, F. Monaco, and M. M. Robertson. 2009. "The behavioral spectrum of Gilles de la Tourette syndrome." *Journal of Neuropsychiatry and Clinical Neurosciences* 21 (1):13–23. doi:10.1176/appi.neuropsych.21.1.13.

Center for Disease Control and Prevention (CDC). 2009. "Prevalence of diagnosed Tourette syndrome in persons aged 6–17 years—United States, 2007." *MMWR Morbidity and Mortality Weekly Report* 58 (21):581–5.

Crossley, E., S. Seri, J. S. Stern, M. M. Robertson, and A. E. Cavanna. 2014. "Premonitory urges for tics in adult patients with Tourette syndrome." *Brain and Development* 36 (1):45–50. doi:10.1016/j.braindev.2012.12.010.

DeCoteau, W. E., C. Thorn, D. J. Gibson, R. Courtemanche, P. Mitra, Y. Kubota, and A. M. Graybiel. 2007. "Oscillations of local field potentials in the rat dorsal striatum during spontaneous and instructed behaviors." *Journal of Neurophysiology* 97 (5):3800–5. doi:10.1152/jn.00108.2007.

Deng, H., K. Gao, and J. Jankovic. 2012. "The genetics of Tourette syndrome." *Nature Reviews Neurology* 8 (4):203–13. doi:10.1038/nrneurol.2012.26.

Draganski, B., D. Martino, A. E. Cavanna, C. Hutton, M. Orth, M. M. Robertson, H. D. Critchley, and R. S. Frackowiak. 2010. "Multispectral brain morphometry in Tourette syndrome persisting into adulthood." *Brain* 133 (Pt 12):3661–75. doi:10.1093/brain /awq300.

Eller, T. 2011. "Deep brain stimulation for Parkinson's disease, essential tremor, and dystonia." *Disease-a-Month: DM* 57 (10):638–46. doi:10.1016/j.disamonth.2011.09.002.

Fries, P. 2005. "A mechanism for cognitive dynamics: Neuronal communication through neuronal coherence." *Trends in Cognitive Science* 9 (10):474–80. doi:10.1016/j.tics.2005.08.011.

Fries, P. 2009. "Neuronal gamma-band synchronization as a fundamental process in cortical computation." *Annual Review of Neuroscience* 32:209–24. doi:10.1146/annurev .neuro.051508.135603.

Ganos, C., V. Roessner, and A. Munchau. 2013. "The functional anatomy of Gilles de la Tourette syndrome." *Neuroscience and Biobehavioral Reviews* 37 (6):1050–62. doi:10.1016/j .neubiorev.2012.11.004.

Goetz, C. G., E. J. Pappert, E. D. Louis, R. Raman, and S. Leurgans. 1999. "Advantages of a modified scoring method for the Rush Video-Based Tic Rating Scale." *Movement Disorders* 14 (3):502–6.

Goodman, W. K., K. D. Foote, B. D. Greenberg, N. Ricciuti, R. Bauer, H. Ward, N. A. Shapira et al. 2010. "Deep brain stimulation for intractable obsessive compulsive disorder: Pilot study using a blinded, staggered-onset design." *Biological Psychiatry* 67 (6):535–42. doi:10.1016/j.biopsych.2009.11.028.

Hariz, M. I. and M. M. Robertson. 2010. "Gilles de la Tourette syndrome and deep brain stimulation." *European Journal of Neuroscience* 32 (7):1128–34. doi:10.1111/j.1460-9568 .2010.07415.x.

Hassler, R. and G. Dieckmann. 1970. "Stereotaxic treatment of tics and inarticulate cries or coprolalia considered as motor obsessional phenomena in Gilles de la Tourette's disease." *Revue Neurologique (Paris)* 123 (2):89–100.

Houeto, J. L., C. Karachi, L. Mallet, B. Pillon, J. Yelnik, V. Mesnage, M. L. Welter, et al. 2005. "Tourette's syndrome and deep brain stimulation." *Journal of Neurology, Neurosurgery and Psychiatry* 76 (7):992–5. doi:10.1136/jnnp.2004.043273.

Hu, W., B. T. Klassen, and M. Stead. 2011. "Surgery for movement disorders." *Journal of Neurosurgical Sciences* 55 (4):305–17.

Hughes, J. R. 2008. "Gamma, fast, and ultrafast waves of the brain: Their relationships with epilepsy and behavior." *Epilepsy & Behavior* 13 (1):25–31. doi:10.1016/j.yebeh.2008.01.011.

Jankovic, J. and R. Kurlan. 2011. "Tourette syndrome: Evolving concepts." *Movement Disorders* 26 (6):1149–56. doi:10.1002/mds.23618.

Kataoka, Y., P. S. Kalanithi, H. Grantz, M. L. Schwartz, C. Saper, J. F. Leckman, and F. M. Vaccarino. 2010. "Decreased number of parvalbumin and cholinergic interneurons in the striatum of individuals with Tourette syndrome." *Journal of Comparative Neurology* 518 (3):277–91. doi:10.1002/cne.22206.

Kurlan, R., P. G. Como, B. Miller, D. Palumbo, C. Deeley, E. M. Andresen, S. Eapen, and M. P. McDermott. 2002. "The behavioral spectrum of tic disorders: A community-based study." *Neurology* 59 (3):414–20.

Leckman, J. F. 2002. "Tourette's syndrome." *Lancet* 360 (9345):1577–86. doi:10.1016/S0140 -6736(02)11526-1.

Leckman, J. F., M. A. Riddle, M. T. Hardin, S. I. Ort, K. L. Swartz, J. Stevenson, and D. J. Cohen. 1989. "The Yale Global Tic Severity Scale: Initial testing of a clinician-rated scale of tic severity." *Journal of the American Academy of Child and Adolescent Psychiatry* 28 (4):566–73. doi:10.1097/00004583-198907000-00015.

References

Leckman, J. F., F. M. Vaccarino, P. S. Kalanithi, and A. Rothenberger. 2006. "Annotation: Tourette syndrome: A relentless drumbeat—Driven by misguided brain oscillations." *Journal of Child Psychology and Psychiatry* 47 (6):537–50. doi:10.1111/j.1469-7610.2006.01620.x.

Leckman, J. F., M. H. Bloch, M. E. Smith, D. Larabi, and M. Hampson. 2010. "Neurobiological substrates of Tourette's disorder." *Journal of Child and Adolescent Psychopharmacology* 20 (4):237–47. doi:10.1089/cap.2009.0118.

Lee, K. H., S. Y. Chang, D. P. Jang, I. Kim, S. Goerss, J. Van Gompel, P. Min et al. 2011. "Emerging techniques for elucidating mechanism of action of deep brain stimulation." *Conference Proceedings: Annual International Conference of the IEEE Engineering in Medicine and Biology Society* 2011:677–80. doi:10.1109/IEMBS.2011.6090152.

Llinas, R. R., U. Ribary, D. Jeanmonod, E. Kronberg, and P. P. Mitra. 1999. "Thalamocortical dysrhythmia: A neurological and neuropsychiatric syndrome characterized by magnetoencephalography." *Proceedings of the National Academy of Sciences of the United States of America* 96 (26):15222–7.

Ludolph, A. G., F. D. Juengling, G. Libal, A. C. Ludolph, J. M. Fegert, and J. Kassubek. 2006. "Grey-matter abnormalities in boys with Tourette syndrome: Magnetic resonance imaging study using optimised voxel-based morphometry." *British Journal of Psychiatry* 188:484–5. doi:10.1192/bjp.bp.105.008813.

Mann, J. M., K. D. Foote, C. W. Garvan, H. H. Fernandez, C. E. Jacobson, R. L. Rodriguez, I. U. Haq et al. 2009. "Brain penetration effects of microelectrodes and DBS leads in STN or GPi." *Journal of Neurology, Neurosurgery and Psychiatry* 80 (7):794–7. doi:10.1136/jnnp.2008.159558.

Marceglia, S., D. Servello, G. Foffani, M. Porta, M. Sassi, S. Mrakic-Sposta, M. Rosa, S. Barbieri, and A. Priori. 2010. "Thalamic single-unit and local field potential activity in Tourette syndrome." *Movement Disorders* 25 (3):300–8. doi:10.1002/mds.22982.

Matsumoto, N., T. Minamimoto, A. M. Graybiel, and M. Kimura. 2001. "Neurons in the thalamic CM-Pf complex supply striatal neurons with information about behaviorally significant sensory events." *Journal of Neurophysiology* 85 (2):960–76.

McNaught, K. S. and J. W. Mink. 2011. "Advances in understanding and treatment of Tourette syndrome." *Nature Reviews Neurology* 7 (12):667–76. doi:10.1038/nrneurol.2011.167.

Miller, A. M., R. Bansal, X. Hao, J. P. Sanchez-Pena, L. J. Sobel, J. Liu, D. Xu et al. 2010. "Enlargement of thalamic nuclei in Tourette syndrome." *Archives of General Psychiatry* 67 (9):955–64. doi:10.1001/archgenpsychiatry.2010.102.

Minamimoto, T. and M. Kimura. 2002. "Participation of the thalamic CM-Pf complex in attentional orienting." *Journal of Neurophysiology* 87 (6):3090–101.

Minamimoto, T., Y. Hori, and M. Kimura. 2009. "Roles of the thalamic CM-PF complex-Basal ganglia circuit in externally driven rebias of action." *Brain Research Bulletin* 78 (2–3):75–9. doi:10.1016/j.brainresbull.2008.08.013.

Mink, J. W. 2001. "Basal ganglia dysfunction in Tourette's syndrome: A new hypothesis." *Pediatric Neurology* 25 (3):190–8.

Molnar, G. F., A. Sailer, C. A. Gunraj, D. I. Cunic, A. E. Lang, A. M. Lozano, E. Moro, and R. Chen. 2005. "Changes in cortical excitability with thalamic deep brain stimulation." *Neurology* 64 (11):1913–19. doi:10.1212/01.WNL.0000163985.89444.DD.

Montgomery, E. B., Jr. and J. T. Gale. 2008. "Mechanisms of action of deep brain stimulation (DBS)." *Neuroscience and Biobehavioral Reviews* 32 (3):388–407. doi:10.1016/j.neubiorev.2007.06.003.

Morrell, M. 2006. "Brain stimulation for epilepsy: Can scheduled or responsive neurostimulation stop seizures?" *Current Opinion in Neurology* 19 (2):164–8. doi:10.1097/01.wco.0000218233.60217.84.

Morrell, M. J. and RNS System in Epilepsy Study Group. 2011. "Responsive cortical stimulation for the treatment of medically intractable partial epilepsy." *Neurology* 77 (13):1295–304. doi:10.1212/WNL.0b013e3182302056.

Okun, M. S., H. H. Fernandez, K. D. Foote, T. K. Murphy, and W. K. Goodman. 2008. "Avoiding deep brain stimulation failures in Tourette syndrome." *Journal of Neurology, Neurosurgery and Psychiatry* 79 (2):111–12. doi:10.1136/jnnp.2007.135715.

Okun, M. S., K. D. Foote, S. S. Wu, H. E. Ward, D. Bowers, R. L. Rodriguez, I. A. Malaty, W. K. Goodman, D. M. Gilbert, H. C. Walker, J. W. Mink, S. Merritt, T. Morishita, J. C. Sanchez. 2013. "A trial of scheduled deep brain stimulation for tourette syndrome: Moving away from continuous deep brain stimulation paradigms." *JAMA Neurology* 70 (1):85–94. doi:10.1001/jamaneurol.2013.580.

Peterson, B. S., P. Thomas, M. J. Kane, L. Scahill, H. Zhang, R. Bronen, R. A. King, J. F. Leckman, and L. Staib. 2003. "Basal ganglia volumes in patients with Gilles de la Tourette syndrome." *Archives of General Psychiatry* 60 (4):415–24. doi:10.1001/archpsyc.60.4.415.

Piedad, J. C., H. E. Rickards, and A. E. Cavanna. 2012. "What patients with Gilles de la Tourette syndrome should be treated with deep brain stimulation and what is the best target?" *Neurosurgery* 71 (1):173–92. doi:10.1227/NEU.0b013e3182535a00.

Plessen, K. J., R. Bansal, and B. S. Peterson. 2009. "Imaging evidence for anatomical disturbances and neuroplastic compensation in persons with Tourette syndrome." *Journal of Psychosomatic Research* 67 (6):559–73. doi:10.1016/j.jpsychores.2009.07.005.

Porta, M., D. Servello, M. Sassi, A. Brambilla, S. Defendi, A. Priori, and M. Robertson. 2009. "Issues related to deep brain stimulation for treatment-refractory Tourette's syndrome." *European Neurology* 62 (5):264–73. doi:10.1159/000235595.

Raeva, S. N. 2006. "The role of the parafascicular complex (CM-Pf) of the human thalamus in the neuronal mechanisms of selective attention." *Neuroscience and Behavioral Physiology* 36 (3):287–95. doi:10.1007/s11055-006-0015-y.

Robertson, M. M. 2012. "The Gilles de la Tourette syndrome: The current status." *Archives of Disease in Childhood. Education and Practice Edition* 97 (5):166–75. doi:10.1136/archdischild-2011-300585.

Robertson, M. M., V. Eapen, and A. E. Cavanna. 2009. "The international prevalence, epidemiology, and clinical phenomenology of Tourette syndrome: A cross-cultural perspective." *Journal of Psychosomatic Research* 67 (6):475–83. doi:10.1016/j.jpsychores.2009.07.010.

Roessner, V., S. Overlack, J. Baudewig, P. Dechent, A. Rothenberger, and G. Helms. 2009. "No brain structure abnormalities in boys with Tourette's syndrome: A voxel-based morphometry study." *Movement Disorders* 24 (16):2398–403. doi:10.1002/mds.22847.

Rosa, M., G. Giannicola, S. Marceglia, M. Fumagalli, S. Barbieri, and A. Priori. 2012. "Neurophysiology of deep brain stimulation." *International Review of Neurobiology* 107:23–55. doi:10.1016/B978-0-12-404706-8.00004-8.

Rouse, A. G., S. R. Stanslaski, P. Cong, R. M. Jensen, P. Afshar, D. Ullestad, R. Gupta, G. F. Molnar, D. W. Moran, and T. J. Denison. 2011. "A chronic generalized bi-directional brain–machine interface." *Journal of Neural Engineering* 8 (3):036018. doi:papers2://publication/doi/10.1088/1741-2560/8/3/036018.

Scharf, J. M., L. L. Miller, C. A. Mathews, and Y. Ben-Shlomo. 2012. "Prevalence of Tourette syndrome and chronic tics in the population-based Avon longitudinal study of parents and children cohort." *Journal of the American Academy of Child and Adolescent Psychiatry* 51 (2):192–201.e5. doi:10.1016/j.jaac.2011.11.004.

Schreckenberger, M., C. Lange-Asschenfeldt, M. Lochmann, K. Mann, T. Siessmeier, H. G. Buchholz, P. Bartenstein, and G. Grunder. 2004. "The thalamus as the generator and modulator of EEG alpha rhythm: A combined PET/EEG study with lorazepam challenge in humans." *Neuroimage* 22 (2):637–44. doi:10.1016/j.neuroimage.2004.01.047.

References

Servello, D., M. Porta, M. Sassi, A. Brambilla, and M. M. Robertson. 2008. "Deep brain stimulation in 18 patients with severe Gilles de la Tourette syndrome refractory to treatment: The surgery and stimulation." *Journal of Neurology, Neurosurgery and Psychiatry* 79 (2):136–42. doi:10.1136/jnnp.2006.104067.

Stanslaski, S., P. Afshar, P. Cong, J. Giftakis, P. Stypulkowski, D. Carlson, D. Linde, D. Ullestad, A.-T. Avestruz, and T. Denison. 2011. "Design and validation of a fully implantable, chronic, closed-loop neuromodulation device with concurrent sensing and stimulation." *IEEE Transactions on Neural Systems and Rehabilitation Engineering* 20 (4):410–21.

Steeves, T. D., J. H. Ko, D. M. Kideckel, P. Rusjan, S. Houle, P. Sandor, A. E. Lang, and A. P. Strafella. 2010. "Extrastriatal dopaminergic dysfunction in Tourette syndrome." *Annals of Neurology* 67 (2):170–81. doi:10.1002/ana.21809.

Steriade, M., D. Contreras, F. Amzica, and I. Timofeev. 1996. "Synchronization of fast (30–40 Hz) spontaneous oscillations in intrathalamic and thalamocortical networks." *Journal of Neuroscience* 16 (8):2788–808.

Steriade, M., R. C. Dossi, D. Pare, and G. Oakson. 1991. "Fast oscillations (20–40 Hz) in thalamocortical systems and their potentiation by mesopontine cholinergic nuclei in the cat." *Proceedings of the National Academy of Sciences of the United States of America* 88 (10):4396–400.

Sudhyadhom, A., I. U. Haq, K. D. Foote, M. S. Okun, and F. J. Bova. 2009. "A high resolution and high contrast MRI for differentiation of subcortical structures for DBS targeting: The Fast Gray Matter Acquisition T1 Inversion Recovery (FGATIR)." *Neuroimage* 47 Suppl 2:T44–52. doi:10.1016/j.neuroimage.2009.04.018.

Sudhyadhom, A., M. S. Okun, K. D. Foote, M. Rahman, and F. J. Bova. 2012. "A three-dimensional deformable brain atlas for DBS targeting. I. Methodology for Atlas creation and artifact reduction." *The Open Neuroimaging Journal* 6:92–8. doi:10.2174/18744400012 06010092.

Sun, F. T., M. J. Morrell, and R. E. Wharen, Jr. 2008. "Responsive cortical stimulation for the treatment of epilepsy." *Neurotherapeutics* 5 (1):68–74. doi:10.1016/j.nurt.2007.10.069.

Temel, Y. and V. Visser-Vandewalle. 2004. "Surgery in Tourette syndrome." *Movement Disorders* 19 (1):3–14. doi:10.1002/mds.10649.

The Tourette Syndrome Classification Study Group. 1993. "Definitions and classification of tic disorders." 1993. *Archives of Neurology* 50 (10):1013–6.

Tourette, G. de la. 1885. "Study of a nervous disorder characterized by motor incoordination with echolalia and coprolalia." *Archives of Neurology* 9 (19–42):158–200.

Uhlhaas, P. J. and W. Singer. 2006. "Neural synchrony in brain disorders: Relevance for cognitive dysfunctions and pathophysiology." *Neuron* 52 (1):155–68. doi:10.1016/j.neuron .2006.09.020.

Uhlhaas, P. J. and W. Singer. 2010. "Abnormal neural oscillations and synchrony in schizophrenia." *Nature Reviews. Neuroscience* 11 (2):100–13. doi:10.1038/nrn2774.

Vertes, R. P., W. B. Hoover, and J. J. Rodriguez. 2012. "Projections of the central medial nucleus of the thalamus in the rat: Node in cortical, striatal and limbic forebrain circuitry." *Neuroscience.* doi:10.1016/j.neuroscience.2012.04.067.

Visser-Vandewalle, V. 2007. "DBS in tourette syndrome: Rationale, current status and future prospects." *Acta Neurochirurgica. Supplement* 97 (Pt 2):215–22.

Visser-Vandewalle, V., L. Ackermans, C. van der Linden, Y. Temel, M. A. Tijssen, K. R. Schruers, P. Nederveen et al. 2006. "Deep brain stimulation in Gilles de la Tourette's syndrome." *Neurosurgery* 58 (3):E590. doi:10.1227/01.NEU.0000207959.53198.D6.

Visser-Vandewalle, V., Y. Temel, Ch van der Linden, L. Ackermans, and E. Beuls. 2004. "Deep brain stimulation in movement disorders. The applications reconsidered." *Acta Neurologica Belgica* 104 (1):33–6.

Vitek, J. L. 2002. "Mechanisms of deep brain stimulation: Excitation or inhibition." *Movement Disorders* 17 Suppl 3:S69–72. doi:10.1002/mds.10144.

Walker, H. C., H. Huang, C. L. Gonzalez, J. E. Bryant, J. Killen, R. C. Knowlton, E. B. Montgomery, Jr. et al. 2012. "Short latency activation of cortex by clinically effective thalamic brain stimulation for tremor." *Movement Disorders* 27 (11):1404–12. doi:10.1002/mds.25137.

Wanderer, S., V. Roessner, R. Freeman, N. Bock, A. Rothenberger, and A. Becker. 2012. "Relationship of obsessive-compulsive disorder to age-related comorbidity in children and adolescents with Tourette syndrome." *Journal of Developmental and Behavioral Pediatrics* 33 (2):124–33. doi:10.1097/DBP.0b013e31823f6933.

Wong, D. F., J. R. Brasic, H. S. Singer, D. J. Schretlen, H. Kuwabara, Y. Zhou, A. Nandi et al. 2008. "Mechanisms of dopaminergic and serotonergic neurotransmission in Tourette syndrome: Clues from an in vivo neurochemistry study with PET." *Neuropsychopharmacology* 33 (6):1239–51. doi:10.1038/sj.npp.1301528.

Xiao, D., B. Zikopoulos, and H. Barbas. 2009. "Laminar and modular organization of prefrontal projections to multiple thalamic nuclei." *Neuroscience* 161 (4):1067–81. doi:10.1016/j.neuroscience.2009.04.034.

Zhuang, P., M. Hallett, X. Zhang, J. Li, Y. Zhang, and Y. Li. 2009. "Neuronal activity in the globus pallidus internus in patients with tics." *Journal of Neurology, Neurosurgery and Psychiatry* 80 (10):1075–81. doi:10.1136/jnnp.2008.161869.

Index

Note: Page numbers ending in "f" refer to figures. Page numbers ending in "t" refer to tables.

Index _____

232